Project Management in Cloud and IoT Applications

Pramod Chandra P. Bhatt
Naresh Kumar Sehgal • Richard Castro

Project Management in Cloud and IoT Applications

Second Edition

Springer

Pramod Chandra P. Bhatt
Pattandur Agrahara, Whitefield
Athashri Apts, A-026
Bangalore, Karnataka, India

Naresh Kumar Sehgal
Securing the Cloud, LLC
Santa Clara, CA, USA

Richard Castro
Rebelstack
Santa Monica, CA, USA

ISBN 978-3-031-99426-5 ISBN 978-3-031-99427-2 (eBook)
https://doi.org/10.1007/978-3-031-99427-2

This Springer imprint is published by the registered company Springer Nature Switzerland AG
The registered company address is: Gewerbestrasse 11, 6330 Cham, Switzerland

Pramod Bhatt would like to dedicate this book to Prof. V. Rajaraman, who encouraged and pioneered research work in the then emerging area of computer science in late 1960s. Prof. Rajaraman's efforts helped to build the formidable reputation IIT Kanpur enjoys as one of the leading institutions in computer science in India and abroad.

Naresh Sehgal would like to dedicate this book to Dr. John M. Acken, his manager at Intel and PhD co-advisor at Syracuse University during the early 90s. John continues to guide and has taught him valuable skills about computer security. Naresh enjoys regular participation in John's research at Portland State University, Oregon.

Richard Castro would like to dedicate this book to the late Dr. Dale A. Baker, who helped him to grow as a scientist and engineer. Dr. Baker paved the way for Richard to a long and fulfilling career in biotechnology.

Foreword for the Second Edition

Cloud has become the backbone of the modern economy. Digital native enterprises are born in the Cloud, while others are increasingly migrating to the Cloud from on-premises infrastructure. As AI becomes more pervasive, the Cloud will become an even more central part of the modern economic infrastructure as almost all key AI technologies are being invented and deployed in the Cloud. As the Cloud has become more critical, it has also evolved. It started with a single dominant player, but today there are many Cloud providers each with their own strengths and multiple paradigms—multi-Cloud, hybrid Cloud, sovereign Cloud—have taken hold.

As such, Cloud expertise is no longer a nice-to-have but rather a must-have for technologists today. Thus, this book fulfills an important function by delving deep into multiple aspects of the modern Cloud—providers, consumers, and business models. It provides a nice overview of project management, software development, DevOps, and other topics that determine the success of building and deploying in the Cloud. Further, by providing additional insights on the Internet of Things (IoT), this book will drive meaningful acceleration in IoT projects.

In the second edition of the book, the authors have also discussed several interesting case studies of companies who have benefited from the Cloud. I believe readers will find this chapter fascinating, and it will trigger many ideas on accelerating the transformation of their workplaces.

DigitalOcean

Digital Ocean
San Francisco, CA, USA
May 24, 2025

Bratin Saha

Foreword for the Second Edition

Foreword for the First Edition

With the invention of technology comes a life cycle of use, starting with the early adopters and eventually accepted by the masses. This move from novel to normal is where we find ourselves today with Cloud computing. With the attributes of rapid deployment, high efficiency, and centralized management, investment in Cloud-based services now consumes IT leaders, and those services have transformed the way businesses large and small think about infrastructure and applications. Cloud Computing has been an accelerant to the pace of technological adoption.

However, with the past decades bringing extensive and expansive Cloud adoption, it has become apparent that the promise of Cloud Computing does not come without calculated planning. Just as CIOs complained in the 1990s of "server spawl," they are now faced with "Cloud sprawl." The learning is the Cloud did not eliminate the need for thoughtful architecture, development, and deployment of IT services. The promised efficiencies in use and cost come only though meaningful project design, beginning with the business need.

This book fills a void with valuable instructions for those looking to the Cloud for IT transformation. It provides practical instruction in Cloud project management across hardware and software. The authors shed light on how to actuate the true benefits of Cloud computing.

NovaSignal

Diane Bryant

CEO and Chairman

www.NovaSignal.com

NovaSignal
Los Angeles, CA, USA
June 8, 2023

Diane Bryant

Preface

This book was conceived when Pramod Bhatt challenged Naresh if there is another book hiding in his heart after we had already co-authored three editions of a book on Cloud Computing with Security and Scalability.[1]

For the second edition, a desire to explore the ever-growing synergy of Cloud and IoT applications was considered. To this end, Richard Castro, who has over 25 years of experience in electronic product design and previously had contributed to the first edition of this book as a reviewer. He was requested to provide his insights on the IoT aspects and its relations to the Cloud.

We reckoned that nowadays software development is Cloud based and happens across geographies. We discussed the idea of capturing this new norm in a book. Both of us have somewhat complimentary backgrounds but share many common experiences. As an academic, Pramod Bhatt has been extensively using and promoting Agile, TDD, and DevOps with his students and the companies that he has consulted. Naresh K. Sehgal, as an engineer had worked for over three decades at Intel and used Agile, DevOps, and managed Cloud-based operations across geographies. We agreed to co-author this book to capture the gist of what we have learnt and assimilated over the several decades of our careers.

There are numerous books and papers on Agile programming, decomposition methods using software engineering methodologies, team dynamics, and project management to execute complex projects. Although there are publications about outsourcing, insourcing, and working with remote teams, there is still a dearth of knowledge on the complexity of managing Cloud software projects that encompass both hardware and software elements and are executed by teams spread across multiple remote geographies. This is not to suggest that such projects do not exist, but the knowledge to execute them has not been codified. It is somewhat of black art depending on the skills of the leaders and program/project managers to ensure that such projects stay on the course, and within the budget to deliver the desired results. This book will review the current state of the art in the management of Cloud-based

[1] https://link.springer.com/book/10.1007/978-3-031-07242-0

software projects and then expose the readers to real-life examples to cull out guidelines and rules that can be then applied to other such projects.

The book begins by asking an innocuous question: what is a project? Then delves into Cloud Project Management (CPM) and how that may differ from other software projects in-house. We will explain how to prepare a business case for doing a CPM. Finance is indeed the horse that drives the cart to deliver any successful project. That is necessary, but clearly not sufficient. Many human elements are needed so we explain the roles and responsibilities for executing a CPM. Next, we dwell upon traditional software development methods, and contrast these with Agile and DevOps. Topics ranging from budgeting, estimating, monitoring, and Service Level Agreements (SLA) are dealt with sufficient details for the practitioners to succeed in delivering a CPM. This book should be quite useful for the students of software engineering, information systems management, and also for the practitioners of Cloud-based project management.

Throughout this book, project management and design considerations with a focus on IoT are presented. An introduction to IoT is followed by more detailed discussions on the IoT product design and support process. The synergistic relationship between IoT products and Cloud applications was always kept in mind. IoT hardware and embedded software issues have been described adding to the usefulness of this edition. This expands the value of this edition to a wider group of readers including students, developers, and practitioners across many disciplines.

A book that covers the practice of executing projects, an ounce of real experience may triumph over tons of theory. With that in mind we include several examples of Cloud-based project case studies in this book. A special focus is placed on the regulations for mobile medical applications. This is because healthcare has become important for the survival of humanity.

We wrap up the book with some rules of thumb for success in executing Cloud-based projects in general and those that are executed across geographies in particular. We do hope you will enjoy and benefit from reading this book, as much as we enjoyed writing it. With many thanks for choosing to read it. We are looking forward to your feedback.[2]

Bangalore, India — Pramod Chandra P. Bhatt
Santa Clara, CA, USA — Naresh Kumar Sehgal
Santa Monica, CA, USA — Richard Castro

[2] nareshksehgal@gmail.com

Acknowledgments

It takes a village of experts to raise a book. The primary project of a Cloud-based application described in this book was architected and implemented by Shiv Shankar. He was the VP of Cloud engineering and leads development of medical applications at NovaSignal.

Richard Castro helped with the review of first edition and joined as a co-author for this book. We are also grateful to Dr. Prasad Ram for writing a section on Personalizing Complex Human Experiences with MyGooru. We appreciate the valuable time of Dr. Bratin Saha, CPTO of Digital Ocean, to write the Preface for the second edition of this book.

Lastly, and more importantly, this book benefited from management support by Diane Bryant. She, as the CEO of NovaSignal, encouraged innovations and acts as a pivot of the company to use Cloud to bring it firmly into the twenty-first century.

Competing Interests The authors have no competing interests to declare that are relevant to the content of this manuscript.

Abstract

Cloud Computing is currently the default mode of IT services delivery and operations. The horizon of Cloud Computing has expanded further with the emergence of IoT, which between them provide end-to-end solutions.

In this edition, we explain the theory and practices of software engineering as applied to successful project execution and management, for real-life Cloud-based products. The readers will find the following topics well covered:

1. Management of projects in general, and Clouds in particular, including budgeting.
2. Best practices in Software engineering, with emphasis on using Agile methodology.
3. IoT products, their deployment, including Cloud-based support and management.
4. Regulatory compliance.
5. New value creation and future opportunities.

This book aims to shine new light on how to deliver Cloud and IoT projects on time and within budget.

Abstract

Contents

About the Authors

Pramod Chandra P. Bhatt started his teaching career in 1965 at IIT Kanpur, and then moved to IIT Delhi in 1969, and retired from IIT Delhi in 1996. Prof. Bhatt also worked as a visiting professor at the University of Ottawa, McGill University, Montreal (Canada), Universities of Dortmund, Paderborn and Bochum (Germany), and Kochi University of Technology (Japan). Prof. Bhatt has an ME from Calcutta University and a PhD from IIT Kanpur. He has also been a Konrad Zuse Fellow at the University of Dortmund.

Naresh Kumar Sehgal worked at NovaSignal Corp for over 3 years as the Sr. VP of Cloud Engineering and CISO (Chief Information Security Officer). Before that he was at Intel for more than 31 years in various engineering and management roles. Naresh has earned a BE (Electrical Engineering) from Punjab Engineering College, MS and PhD from Syracuse University in Computer Engineering. Naresh taught Cloud Computing at Santa Clara University, where he also earned an MBA. He currently consults on Cloud Computing and Product Management tasks.

Richard Castro has over 20 years of product development design, leadership, and entrepreneurial experience in medical device and safety critical markets. Specializing in embedded, RF, and high-speed digital electronics, Richard has led many small to mid-size engineering teams which successfully brought FDA-approved products to the market.

Abbreviations

ACID	Atomicity, consistency, isolation, and durability
AI	Artificial intelligence
ANSI	American National Standards Institute
API	Application programming interface
ASME	American Society of Mechanical Engineers
ASTM	American Society for Testing and Materials
AWS	Amazon Web Services
BDD	Behavior-driven development
BE	Broadband engine
BLE	Bluetooth low energy
CCC	Collaborative Cancer Cloud
CCPA	California Consumer Privacy Act
CCS	Critical chain scheduling
CDN	Content delivery network
CEO	Chief executive officer
CICD	Continuous integration and continuous delivery
CISO	Chief Information Security Officer
CPM	Cloud project management
CPU	Central processing unit
CSP	Cloud service provider
CTO	Chief technology officer
DDOS	Distributed denial of service
DevOps	Development and operations
DOS	Denial of service
EC2	Elastic Compute Cloud
ECC	Error correcting codes
ECO	Engineering change order
EdTech	Education technology
EMR	Electronics medical records
ERD	Entity relationship diagram
ERP	Enterprise resource planning

FCC	Federal Communications Commission
FDA	Food and Drug Administration
FDI	Foreign Direct Investment
FL	Federated learning
FPGA	Field-Programmable Gate Array
GB	Giga bytes
GCP	Google Cloud Platform
GDP	Gross domestic product
GDPR	General Data Protection Regulation
GPU	Graphics processing unit
HA	High availability
HIPAA	Health Insurance Portability And Accountability Act
HR	Human resources
HTTP	Hypertext Transfer Protocol
HTTPS	Hypertext Transfer Protocol Secure
I/O	Input/Output
IaaS	Infrastructure as a Service
IAM	Identity and Access Management
IBM	International Business Machines
ICD	Interface Control Document
IEC	International Electrotechnical Commission
IEEE	Institution of Electrical and Electronics Engineers
INR	Indian rupees
IoT	Internet of Things
IP	Internet Protocol
ISO	International Standards Organization
IT	Information Technology
ITPM	Information Technology Project Management
KLOC	Kilo Lines Of Code
LLM	Large language model
LoC	Lines of Code
ML	Machine Learning
MRD	Marketing requirements document
MWL	Modality work list
NIST	National Institute of Science and Technology
NSA	National Security Agency
OO	Object oriented
OS	Operating system
PaaS	Platform as a Service
PACS	Picture Archiving and Communication System
PCB	Printed Circuit Board
PERT	Program Evaluation and Review Technique
PHI	Private/Protected/Patient Health Information
PII	Personally Identifiable Information
PM	Program/Project/Product Manager

PMBOK	Project Management Body of Knowledge
PMI	Project Management Institute
PMIC	Power management integrated circuit
PoC	Proof of Concept
PRD	Product requirements document
PWA	Progressive Web App
QA	Quality assurance
ROI	Return on investment
RTF	Realized-tested-features
SaaS	Software as a Service
SaMD	Software as a Medical Device
SDLC	Software Development Life Cycle
SEI	Software Engineering Institute
SLA	Service level agreement
SLM	Small language model
SME	Subject matter expert
SOA	Service-oriented architecture
SoC	System on Chip
SOUP	Software of unknown providence
SRS	System requirements specifications
SSO	Single Sign On
SW	Software
TBO	Total benefits of ownership
TCO	Total cost of ownership
TDD	Test-driven development
TLS	Transport Layer Security
UI	User interface
UID	Unique Identification Number
UL	Underwriters Laboratories
UML	Unified Modeling Language
URL	Unified Resource Locator
USB	Universal Serial Bus
USD	United States Dollar
VM	Virtual machine
VNA	Vendor neutral archive
VPN	Virtual Private Network
WAN	Wide area network
Wi-Fi	Wireless Fidelity
XP	eXtreme Programming
YoY	Year over Year

Chapter 1
Cloud Project Management

1.1 Introduction

Traditionally, people associate the terms like "project" and "project management" with some engineering activity. The project management, as an area of study, is also considered to be within the ambit of engineering discipline. Engineering disciplines have evolved over centuries and so has project management. In comparison, software development and software engineering have evolved only in the last few decades. Furthermore, Cloud Computing came in vogue at end of the first decade of the twenty-first century [1]. Most non-IT engineering projects manifest in an artifact that has form and feel which are physical in nature. Software projects and products do not quite fit in that genre. This makes the task of Cloud-based software project management both interesting and challenging. Implementing software that executes in Cloud to provide computational and service support adds to these challenges, such as selection of Cloud vendor, scheduling timely releases and software patch management, ensure minimal security levels and scalability without losing much on performance while dealing with environment-related latencies, etc.

Every project worth doing must have a business case and a statement of purpose. The statement of purpose is duly supported by a strategic plan and a project charter. The show begins once all the mandatory project approvals have been obtained. To begin with, a project manager is appointed. The project manager and the project team have the responsibility of executing the project.

This book has a limited objective to offer an exposition on software project management in the Cloud. Cloud Project Management (CPM) covers software project planning, preparation of project charter, preparing cost and time estimates, and evolving a quality regime strategy to ensure quality of the product or service to be delivered in the Cloud. Every project entails a closure protocol and also must have an evaluation conducted. The evaluation leaves an organization richer in experience and wiser for undertaking more ambitious projects in future. In addition, sometimes it may lead to generating new intellectual property of long-lasting value.

P. C. P. Bhatt et al., *Project Management in Cloud and IoT Applications*,
https://doi.org/10.1007/978-3-031-99427-2_1

Project execution may be managed in-house or may be outsourced. In the last few decades, the momentum of economic globalization has resulted in outsourcing beyond national boundaries. This phenomenon is most dominant in the information technology sector. Software project outsourcing throws up many interesting challenges in project management. We shall explore this aspect as we go along.

In this chapter, our goal will be limited to defining a software project and give some basic flavors of Cloud [1]. Even though we shall build from the basics of the traditional view of the project management, we shall distinguish SW project management from engineering project management and its life cycle.

1.2 Introduction to Internet of Things (IoT)

An Internet of Things (IoT) project stands out as distinct from a solely Cloud-based or a standalone software project. In fact, IoT projects can sometimes be considered in tandem with a Cloud project within the traditional hardware engineering arena. The "Things" in IoT devices are physical devices with electronics embedded in them, typically as hardware sensors and actuators to measure and interact with the real world. While most IoT devices today are relatively simple products such as fitness trackers or smart light bulbs, the complexity and application space of IoT have been expanding to include industrial and safety critical applications.

The convergence of embedded processing power and memory storage improvements, wireless network and protocol expansion, and machine learning research has enabled IoT devices to exhibit new levels of intelligence [2]. For example, single board computers or edge devices can now be coupled with cameras in an IoT device to implement face detection in real time. The IoT device can be programmed to make decisions based on facial recognition. Applications for this technology range from home security to corporate building access management to country border control.

In the industrial application space, IoT devices have the advantage of reducing human error, increasing productivity, and providing safeguards against workplace accidents. One example is the use of smart glasses in manufacturing settings. In a factory, these can be a pair of safety glasses that are connected to the company's internal network and equipped with a heads-up display that can help workers interpret work instructions, detect parts, provide instant feedback on part installation, and notify workers if they are too close to dangerous machinery. This is all done while allowing the worker full use of their hands for assembly.

The above examples provide some insight into the large field of applications available to IoT. Figure 1.1 illustrates many more applications to which IoT devices can be applied. Given the expansive application space, IoT project management—and its relation to Cloud project management—is a valuable skill to foster for any company or venture considering development of an IoT device.

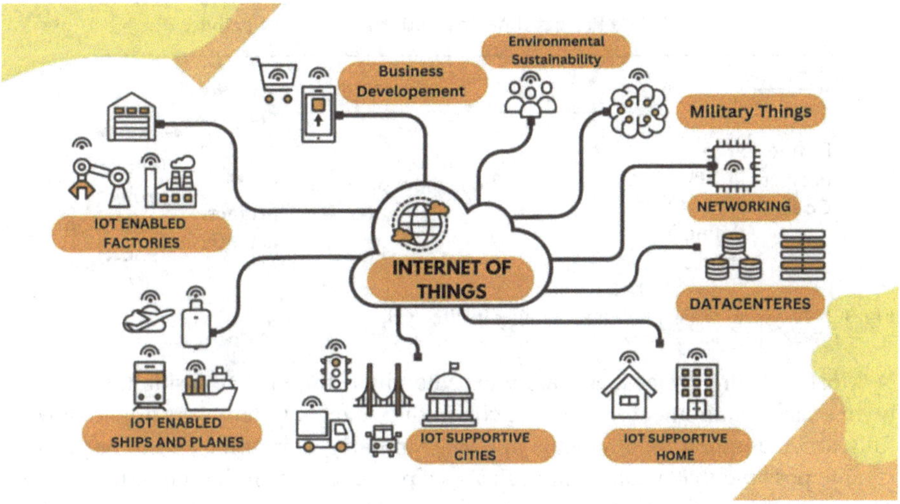

Fig. 1.1 Examples of IoT devices [3]

1.3 What Is a Project?

People often easily relate to the term "project" with some physical manifestation of structural grandeur. Technically, the notion of a project and project management is nearly as old as human civilization. Structures such as Red Fort in Delhi or Eiffel Tower in Paris were massive projects for the eras they were undertaken. Certainly, these projects took many years to complete. Note each was created with a certain goal in mind. In the case of Red Fort, it was to define the seat of Mogul power in India. In the case of the Eiffel tower, it was to mark the centenary of French revolution. Eiffel tower is a monumental Iron insignia of freedom and was created through an open competition. Every such endeavor referred to as "a project" needs planning, mobilization of resources, and execution skills that require close monitoring to achieve the goals. Let us now define a project in general terms.

The Project Management Institute (PMI) defines [4] a "project" as a "temporary endeavor undertaken to accomplish a product, service or result." Note that the word "temporary" requires that project, as an activity, shall have a temporal point for "*beginning*" and an identifiable "*end*." Further, it must result in creating some value for the promoter. In the case of business, a project must offer some *business value* and in the case of societal initiatives, the *value translates into public good.*

Projects are undertaken by people to accomplish certain defined goals. The project activities span a timeline as shown in Fig. 1.2. The project starts with the definition of project goals. Following it, a project plan is prepared, and a project charter is created. The project team takes over the project implementation till it is finished. On completion, a proper closure of the project is performed, and an evaluation report is created. This time marked pattern of activities defines a project's life cycle.

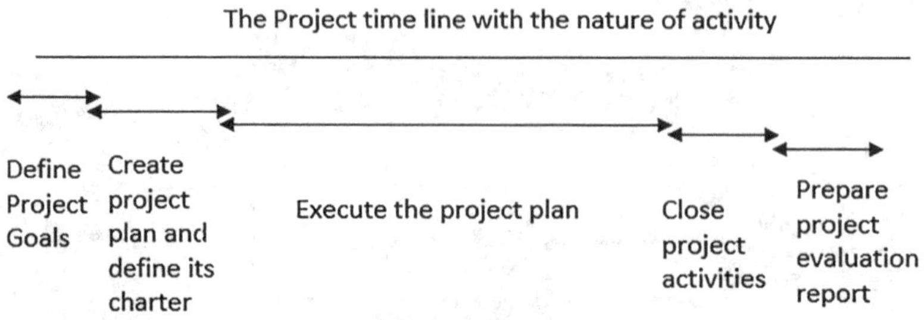

Fig. 1.2 A typical project timeline defining its life cycle

Most software projects are rarely executed in a straight-line fashion, as inevitably the user requirements or market conditions change. Often users are not able to fully articulate their requirements upfront. Similarly, during the execution stage, it is quite possible that a competitor releases a product with similar or better features. This may reduce the market success chances of the currently defined project, so requirements will need to be modified or new requirement have to be added. Nowadays, SW projects are invariably implemented by first creating a prototype. When users see a prototype, new ideas may emerge. This necessitates an iteration to adopt the new requirements by a corresponding change in the SW architecture and/or implementation details. Sometimes, these changes will result in a schedule slip. In that case, the project manager has an option to incrementally accept new requirements in the existing project or defer them to the new version.

The project management discipline has its own vocabulary and definitions. In studying the Cloud-based project management, we stand to benefit by exploring this existing knowledge. The Project Management Institute (PMI) has embodied it in "project management body of knowledge" (PMBOK) [4]. The PMBOK has identified key areas of concern. As a result, the discipline has evolved around the following nine key areas (defined in PMBOK):

1. **Scope management**: A project scope provides a set of *unambiguous definitions that cover all the activities* within a project. The accuracy in the definitions helps to determine if a project achieved what it was set out to accomplish. The scope evokes a true and complete response to the question*: Did the project execution happen as planned?*
2. **Time management**: Based on an estimate and the likely effort required to complete, a project may be decomposed into multiple phases. This is done to ensure that the project is completed as planned, within budget and timeframe. A time schedule needs to be prepared for each phase in the project. *The time management entails instituting a monitoring regime* with the objective to regulate the project activities. In fact, activities may be regulated within each phase to ensure scheduled completion of each phase and eventually of the project itself. Time management is also referred sometimes as *schedule management.*
3. **Cost management**: Every enterprise that engages in implementing a project has to provision a budget to be able to support the people and also ensure timely

supply of resources. Primarily, the cost is estimated based on an assessment of the effort involved and resources required to complete the project. The cost estimates may (almost always) differ from the eventual cost. In addition, every budget provisioning is tricky considering the economic uncertainties. The cost management is essentially a *budgetary exercise to ensure that the project activities do not suffer monetary starvation.*

4. **Quality management**: A quality regime needs to be set up for the final product. The quality assurance requires continuous monitoring and testing to ensure that the quality parameters of the final product are met. Projects may sometimes need supplies from parties that produce specialized components. In practice, the desired quality and cost are often in conflict. One, therefore, needs to ensure that the *desired quality standards are not compromised while staying within the cost.*
5. **Risk management**: In implementing projects, one frequently encounters conflicting demands that require decision to be made with associated risks. In most cases, *the balancing act involves determining pros and cons of choices* made. One needs to ensure that the project is not only secured but, at the end, benefits accrue.
6. **Human resource (HR) management**: Project implementation is a team effort. Human resource management *involves building a project team* with a leader—a project manager—having vision and drive. The key to HR management lies in maintaining high motivational levels of the project team to accomplish project goals.
7. **Communication management**: It is important that there is *accurate reporting on the progress made for the activities within each phase* of the project. The communication is the key to the preparedness, continuity, and hand over between different phases of the project.
8. **Integration management**: Every project shall have multiple activities. Integration management addresses the concerns of *upstream activities feeding into downstream activities.* Sometimes upstream activities dovetail into others taking them forwards.
9. **Procurement management**: Every project spans over a period and has multiple phases. Each phase has its own resource requirement—be it human or physical facilities. The *procurement management deals with the practices that ensure provisioning the required resources* as project transitions from one phase to the other.

A Few Other Observations It should be noted that there are *stakeholders*, and stakeholder management may be an additional consideration. Stakeholders may be within the ambit of the project or others who are not part of the project team or even the enterprise, but they could influence the overall outlook. Broader issues with socio-political impact shall not be included here. However, a conscientious and curious reader should explore this aspect from other sources [5]. This is important in the context of social media and AI or its variant such as Artificial General Intelligence (AGI). However, this book has a limited goal of exploring the dynamics of Software project development and management.

With this general background on projects—we should like to observe that nearly all the management underpinnings captured above apply to software project management as well. Technically, software project management also requires management of people, scope, resources, time, cost, communication, integration, quality, and risk. The rest of the discussion veers around the Cloud-based Information Technology Project Management (ITPM).

1.4 Cloud Project Management

Along the lines of PMBOK, the ITPM defines a project to have a timeline shown in Fig. 1.3. The main difference lies in the third phase where the ITPM stipulates selection of a software implementation model from among many that are possible. Also, note that the last phase places considerable emphasis on identification of the software assets generated. The software assets enrich the repository of the intellectual property of an organization that may accumulate over time. The software assets help in the formulation of future projects. Considerable cost benefits accrue from the reuse of tools and the techniques learned.

In the case of Cloud projects, teams are created with project managers, software architects, developers, programmers, and quality assurance groups. We sometimes learn from media about projects that actually fail. It turns out that the choice of software development life cycle (SDLC) is very critical in determining success of SW projects. We discuss the choice of SDLC later in this book. For now, we would reckon a SW project to be successful if it fulfills the following conditions:

(a) It leads to a software product or creates a unique software service opportunity.
(b) It is completed on-time (or with acceptable time overrun).
(c) It is within the budget (or within acceptable cost overrun).
(d) It meets certain defined business goals through value addition.
(e) Meets appropriate quality parameters and standards.

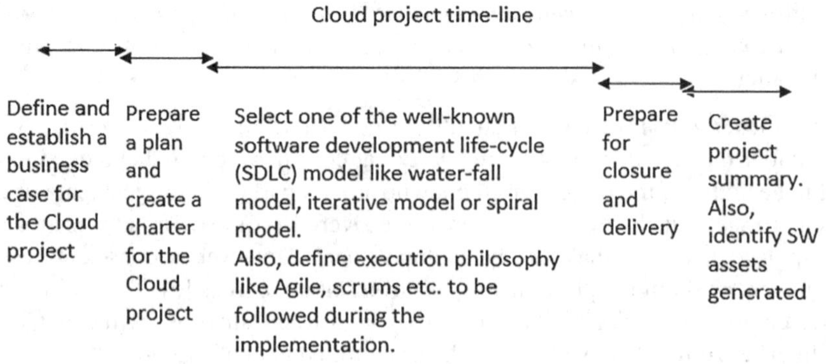

Fig. 1.3 Information technology project management (ITPM) defined project life cycle

(f) Makes the software enterprise richer in experience to undertake more ambitious projects.

Points a–f above have one primary underpinning: *A SW project development must yield an executable code that provides service of value*. Every other consideration on deliverables becomes secondary.

It has been variously argued that SW projects differ from the engineering projects in significant ways. Jack Marchekwa [6] in his 2003 edition (pp 64) had cited the differences enunciated by Dr. Kapoor. Even though some of those are not relevant any more, one still can cite the following distinctive differences:

(a) Engineering projects have physical manifestation as they evolve. Also, the engineers use drawings to offer fairly comprehensible visualizations. Software projects rarely have such physical manifestations. To overcome this drawback, many organizations create a model using rapid prototype [5] to give a look and feel of the final outcome. Simulated executions too come in handy in this regard.
(b) The engineering standards have evolved over time and stabilized. As a consequence, the components used in assembly have a guaranteed assurance of quality and interoperability. The software components and protocols that offer connectivity are still evolving. In many cases, standards and protocols do not even exist or are still emerging. A case in point is IoT (Internet of Things) products using edge computing duly supported by operations in Cloud. For Cloud-based software development, a few standards such as APIs for browsers, third-party integration, and Internet gateways for third-party services exist [1]. However, the quality of some implementations may be suspect. In the projects integrating open-source components is often termed as Software of Unknown Providence (SOUP). The FDA, a US government agency, defines it as the software that is already developed and widely available, but for which adequate records of the development process are not available. Therefore, some extra care must be taken during the integration testing to ensure that original assumptions of the SW component providers remain valid in the new project. Many, if not all, third party components are proprietary. These are available as "under license to use" binaries. Hence, one may not be able to even modify these for the particular context of use. That creeps in risks.
(c) The processes that lead to implementation of engineering projects are well understood. The cost and time overruns are only sometimes out of control. Also, very few engineering projects are aborted. In the case of software, estimates are guesses. In the initial years, the time and cost overruns of 200% were not uncommon. However, since mid-nineties, the software cost and time estimates have become better—as we will show later in this book. But aborting projects mid-way is pretty common. In fact, it approximates to 26% — which is still quite alarming.
(d) Another aspect, not touched by Dr. Kapoor, but quite important is the distributed SW project management. With the extent of outsourcing that happens in the IT sector, a much tighter control and inspection regime needs to be enforced for SW development projects. The global day-time differences, progress in

communications, and collaborative platforms allow for the day-to-day project work to be shared by the work force that may be geographically distant with follow-the-moon strategy for time and energy saving.

(e) Let us compare SW construction with manufacturing activity. Again, we find glaring differences. For instance, the manufacturing process is repeatable, has a quality control-related yield, and all the productivity indices are relatively easily defined. Added to that, the technology gestation is long. For example, a successful automobile design may last a decade or more. Whereas for SW, technology gestation is low, and quality control is essentially design centric and coder capability related. The activity is rarely repeatable! Its design centricity makes SW hugely people dependent, i.e., dependent upon designers and programmers who design, code, and test.

This short comparison of the engineering projects with SW projects provides us enough motivation to undertake SW project management as a discipline in its own right. Note that software nowadays touches nearly every form of human activity—be it social networking, communications, education and health delivery, sports, entertainment, e-governance (including administrative services to the citizen like driving license, land registration, etc.), streaming services for sports, movies and videos on demand, business, financial services like banking, and insurance applications based on Cloud. In other words, more and more IT-related services supported by Cloud-based operations are likely to be used in times to come. This only underlines the need to understand the full dynamics of Cloud project development and its management.

According to the Gartner, worldwide IT spending is projected to total $5.1 trillion in 2024, an increase of 8% from 2023. The software and IT service segments will both see double-digit growth in 2024, largely driven by cloud spending. Global spending on public cloud services is forecast to increase by 20.4% in 2024, and similarly to 2023, the source of growth will be a combination of cloud vendor price increase and increased utilization.

1.5 Software Project Management Statistics

In the 1990, frequent SW project failures were common knowledge. These failures were in terms of aborted projects, unusual time, and cost overruns with low quality in the tow, prompted Standish group to study management of SW projects. This resulted in the famous CHAOS study. CHAOS is quoted very often to build a rationale for SW project management.

The second report produced nearly a decade later is more interesting from our stand point. It reveals that with the experience gained in executing projects and with better understating of project management techniques, the failure rates were brought down, and in addition, the cost and time overruns were better regulated and quality delivered (in terms of defects per kilo lines of code (KLOC) was much lower).

1.5.1 The CHAOS Study

Chaos study was originally conducted in 1995 [7]. It was revised in 2014 generating the 21st edition [8]. To get an insight into the management aspects, the group categorized the companies as large, medium, and small, as shown in Table 1.1.

The large projects handled by large companies were over 2 million USD in cost terms while the medium-sized projects were about 1 million USD and small were less than 0.5 million USD. Further, the projects were categorized as successful, challenged and impaired. The definition of successful being: on-time and within budget. The challenged ones not only had cost and time over runs but lacked in features as well, i.e., these projects were short on meeting stipulated features required. The impaired ones had to be aborted as pulling them through was not feasible.

In Table 1.1, we summarize their findings with the figures indicating average percentages. Table 1.1 shows that nearly 91% of large projects, on an average, had cost and time overruns and did not deliver on the stipulated features. With nearly 84% as the figure for the medium-sized projects, the picture does not appear any rosier. In fact, only 28% of the small projects were successful. This certainly compares very poorly with engineering projects!!

We take another look at these projects for the extent of cost and time runs. The poor record in respect of cost, time, and features and functions is shown in Table 1.2 (in percentage terms). Larger projects have close to 200% time and cost overruns. The smaller projects in fact seem to be worse!!

The interesting point to note is that almost every category underestimated the effort required. The smaller the company, more optimistic it was. Note the figures worsen as the company size shrinks. Also, note that the larger companies failed in delivering the features more often. For non-SW projects, one may be able to get relevant data from PMI [4].

In addition to the two tables, the CHAOS study analyzed SW projects for success and failure. Note CHAOS defined the categories as successful, challenged and

Table 1.1 The CHAOS study on success rate of projects

Company size	% successful projects	% challenged projects	Impaired projects
Large (>2 million USD)	9	61.5	29.5
Medium (~1 million USD)	16.2	46.7	37.1
Small (<0.5 million USD)	28	50.4	21.6

Table 1.2 More from CHAOS study on success rate of projects

Company size	% cost over run	% time over run	% features and functions delivered
Large (>2 million USD)	178	230	42
Medium (~1 million USD)	182	202	65
Small (<0.5 million USD)	214	239	74

Table 1.3 Ten most important reasons for failure or success of SW projects

Rank	Successful	Challenged	Impaired
1	Adequate user involvement	User input inadequate	Incomplete requirement
2	Good executive management support	Incomplete requirement specifications	Lack of user participation
3	Clear requirement specification	Frequent changes in requirements	Inadequate resources
4	Good plan	Poor executive mgmt. support	Expectation not realistic
5	Setting correct level of expectations	Poor understanding of the technology to be used	Lack of executive management support
6	Well-defined shorter-term milestones	Inadequate resources	Frequent changes in requirements
7	Technically well-equipped manpower	Expectation not realistic	Poor planning
8	Motivated ownership	Objectives not well defined	The project was not needed any more
9	Crisp and clear vision statement	Time schedules not realistic	Poor SW management
10	Motivated team with focus	Lack of familiarity with the new technology	Poor understanding of technological issues

impaired. Table 1.3 summarizes the findings. Table 1.3 ranks the reasons from 1 to 10 with 1 having the greatest impact on success or lack of it.

Readers may explore Table 1.3 carefully as it has pointers embedded in it to define when to consider a project successful while also pointing to the reasons for failure. The following key points emanate from the table:

1. **Clearer statement of requirements**: The requirements ought to be better and clearly understood with deeper user involvement. Correct statements about requirements set appropriate expectations. Frequent changes in requirement only add to the effort and add not only to cost and time overruns but also lead to diminishing enthusiasm of people during the project's lifetime.
2. **User participation and setting correct level of expectations**: Related to the requirements is the involvement of the key stakeholder—the user. With user involvement, not only one has a better handle on evolving requirements but also ensuring that the expectations are set at the correct level. In fact, this is the single most reason for the current popularity of Agile processes [9]. We discuss Agile processes in Chap. 3.
3. **Resource planning and technology support**: This is a key SW management issue. One can set better shorter-term milestones when there is a clearer understanding of technology to use and resource management. In that case, fewer things would go wrong. Also, it prepares the team to deliver on the expectations.
4. **Executive management support**: Project team's zeal and motivation suffer in the absence of the executive management support.

There are other ways to churn the data in Table 1.3. For instance, weights may be associated with rank and then analyze the causes of failure (across the projects that are challenged or impaired). Regardless of the data churn, the abovementioned four observations will still stand out in the order—requirements and user participation contribute about 40–50% and SW project management about 40%. Come to think of it, even the management of expectations and management of technology are also required for SW project management.

Carrying forward from CHAOS, and analyzing the Standish Group's later work, Prof. Marchekwa [6] has noted significant improvement in arresting SW project failures. Table 1.4 shows how the successful and impaired project percentages look like. The remaining projects should be considered as challenges.

There are two well recognized project development processes, namely, e.g., Waterfall and Agile processes. Since there are more than 10,000 software projects in the CHAOS study [8], the results in Table 1.5 show that Agile projects have almost 4x success in comparison to Waterfall model. Besides Waterfall projects have 3x the failure rate of Agile projects. If these are results are decomposed by the project sizes, similar differences persist except of small size projects. A reason may be that small projects typically have small teams and shorter execution timelines, so differences between the outcome Agile and Waterfall processes are fewer.

Before leaving this section, one would like to refer to a classic and oft quoted book by Fredrick Brook, "The mythical man-month" [10]. It describes a project that led to the development of a very successful system—IBM 360. The book offers very valuable insights on the dynamics of software projects and offers advice on how complexity may be managed. One of the observations in the book is now reckoned to be Brook's law. It states, "Adding manpower to a late project makes it later." This law seems to hold good most of the time—reinforcing the belief that SW project management is different from engineering project management in many ways.

Table 1.4 Successful and impaired projects from 2011 to 2015 timeframe [8]

MODERN RESOLUTION FOR ALL PROJECTS

	2011	2012	2013	2014	2015
SUCCESSFUL	29%	27%	31%	28%	29%
CHALLENGED	49%	56%	50%	55%	52%
FAILED	22%	17%	19%	17%	19%

The Modern Resolution (OnTime, OnBudget, with a satisfactory result) of all software projects from FY2011–2015 within the new CHAOS database. Please note that for the rest of this report CHAOS Resolution will refer to the Modern Resolution definition not the Traditional Resolution definition.

Table 1.5 Comparison of Agile vs. waterfall process [8]

CHAOS RESOLUTION BY AGILE VERSUS WATERFALL

SIZE	METHOD	SUCCESSFUL	CHALLENGED	FAILED
All Size Projects	Agile	39%	52%	9%
	Waterfall	11%	60%	29%
Large Size Projects	Agile	18%	59%	23%
	Waterfall	3%	55%	42%
Medium Size Projects	Agile	27%	62%	11%
	Waterfall	7%	68%	25%
Small Size Projects	Agile	58%	38%	4%
	Waterfall	44%	45%	11%

The resolution of all software projects from FY2011–2015 within the new CHAOS database, segmented by the agile process and waterfall method. The total number of software projects is over 10,000.

1.5.2 The Upstream-Downstream Effect

Basically, we notice that SW projects have fairly high failure rates, cost, and time overruns and feature fulfillment is also poor. Same extends to Cloud-based Software projects too. This can be seen with the following three different perspectives.

1. **Management perspective:** As a pure management exercise, it points to the lack of *efficiency* and *effectiveness* in the management of projects. Peter Drucker, quoted by Marchekwa [6], defines efficiency as *doing the right thing* and effectiveness as *doing the thing right*. One clearly emphasizes control of wasted effort and the other emphasizes the use or correct means to achieve the desired effect.
2. **Software engineering perspective:** The software engineering community too has been seized of the poor feature fulfillment, cost and schedule overruns. Cost and schedule overruns are often due to *rework* (wasted effort above) required. Rework is required when someone figures out that a certain client or user requirement is not fulfilled. The software engineering community, therefore, invokes the notions of validation and verification through a test regime at every stage of the development. The validation is to check if the program meets the requirement, i.e., *it is the right thing*. Verification ensures that the intent is interpreted

correctly—at every stage of the development activity has been *done the right way.* In this respect, the Agile development and scrum methodologies help in project management.

3. **General engineering perspectives:** Software discipline is young and rapidly evolving with Cloud-based operations appearing on the horizon. Whereas, the tools and quality checks for other disciplines have had more time to evolve. Having said that the automation adoption has been slower in other general engineering disciplines. Basically, some standards and best practices have to evolve to a level of maturity.

The question now to pose is: Notwithstanding our understanding of the management and software engineering issues why do frequent failures happen? Next, we briefly indicate how it happens and how it impacts the SW project and development activity. Also, how does it translate in to cost and schedule over runs. Needless to say that this is indeed a strong and highly motivating challenge in SW project development and its management activity.

Note that much of the *activity within SW projects is performed by people.* These people are knowledge workers with roles like SW architects, designers, programmers, testing specialist, security system experts, middleware experts, UI designers, performance experts, etc. Also, there are folks who install packages and / or maintain them. They experience enormous mutual dependencies. They also experience *cognitive dissonance* which, to quote Wiki [11] manifests as follows: "*Despite contrary evidence, people are biased to think of their choices as correct.*"

In SW projects development activity, people have to use their abilities to deliver a safely executing code. In addition to the cognitive dissonance, *variance in the capabilities of persons* (people-ware) is very high—which leads to variations in expectations. SW project development must withstand this variation in various stages of SW development. So let us first look at a typical development cycle.

Most often program development evolves through multiple stages like requirements analysis, design, coding, testing, and installation. Different persons work in each of these phases. The downstream workers must correctly interpret the upstream person's intent. Cognitive dissonance often leads to misinterpretation. The errors may creep in at any stage. If detected, within that stage it is at best an *error.* Its manifestation later marks it as a *defect*, i.e., *an upstream error is a downstream defect.* Often late detection of defects leads to impaired projects. SW project management needs to take a call on risks in letting which of the defects propagate when the project is finally delivered.

In an article [12], Steve McConnell discusses cost of downstream defect removal for upstream decisions. Figure 1.4 shows how the costs of correcting a defect mount over time.

Let us absorb the message in Fig. 1.4: The cost is usually in terms of the effort which translates to man-months lost due to rework involved. Add to this the penalties due to the delays. So, even while the costs are not quantified in the figure, the following points are evident:

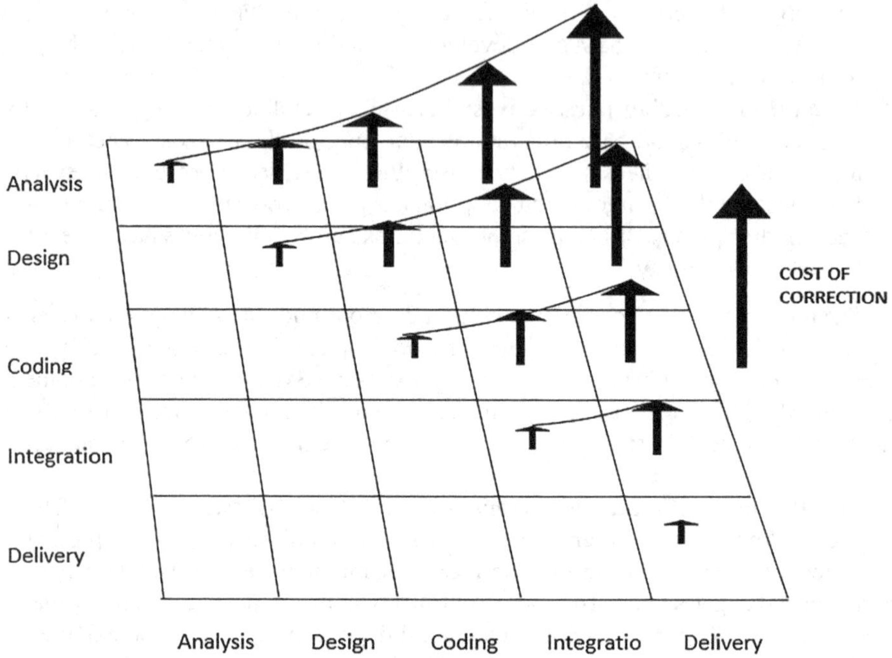

Fig. 1.4 Cost of downstream defect removal

1. The cost of correcting an error within a certain stage is minimal. Let us consider it as x. The error becomes a defect if it permeates into subsequent stages. The costs for defect removal escalate very rapidly in later stages, i.e., it is y times x, where $y \gg 1$.
2. It has been variously estimated that a requirements error manifesting as a defect at the end of project may require over 100 times the effort [13] and, therefore, in cost terms becomes prohibitive.

The discussion provides us a strong motivation to study communications, documentation, inspection and test regimes, defect, and associated risk management in the SW development projects. This is true more so for the Cloud-based projects that get executed across geographies.

An Internet of Things (IoT) project stands out as distinct from a project that is solely Cloud based. In fact, an Internet of Things project can be thought of as an extension of a Cloud project into the traditional hardware engineering arena. The "Things" of IoT devices are physical electronics devices, typically including various hardware sensors and actuators to measure and interact with the physical world. While most IoT devices today are relatively simple products such as fitness trackers or smart light bulbs, the complexity and application space of IoT devices are expanding to include industrial and safety critical applications.

1.6 Constraints in Cloud-Based Software Project Management

Every project has some in-built constraints. Management professionals identify these as the budget, scope, and schedule. Invariably, the project goals set the scope but the schedule and budget may require a tight rope walk [14]. This holds true for Cloud SW project and development activity as well. Additional constraints include deployment cost management, e.g., choice of a Cloud vendor, selection of a data-center location to minimize customer latency, what type of resources to rent in a Public Cloud to ensure the right balance between performance and recurring cost, etc. Figure 1.5 depicts the project cost triangle. More features will need more resources and time. Note that more of every item expands the area of the triangle in Fig. 1.5. The expansion in area translates expanding the project cost. In short, this stipulates enhanced budgetary and resource provisions to implement more features. In most practical situations, however, one always falls short on the budget and resources. Therefore, one needs to have a method for cost and schedule estimates.

The provisioning of resources impacts the project in many ways. For instance, it impacts the team's ability to enhance the scope in terms of features, when and if, new requirements are thrown in as the project progresses. A realistic schedule ensures that there is enough time available not only for implementation but also for testing and putting a defect tracking system. It also allows the team to generate a good documentation—which in turn also enriches the enterprise assets.

We observed earlier that SW projects are deeply people centric. A project team remains motivated when the project provides them with some tangible professional value. The people are the biggest asset an IT enterprise can have. Therefore, project *ownership* is crucial for the success of a project. Enterprises need to partake with benefits to provide incentives to stakeholders including the project team. This brings us to a point where we need to describe roles and responsibilities in the context of project management.

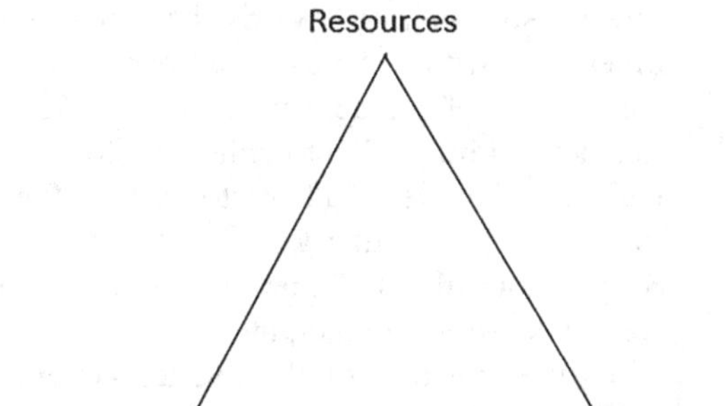

Fig. 1.5 The constraints for implementation of SW project development

1.7 Cloud and IoT Project Roles and Responsibilities

Conway's law [15] captures similarities between an organization's structure and the systems it is required to design. In 1967, a computer programmer Melvin Conway stated the following:

> Organizations which design systems are constrained to produce designs which are copies of the communication structures of these organizations.
> —*Melvin E. Conway, How Do Committees Invent?*

This law is based on the logic that for a product to function, its authors and designers must communicate with each other to ensure compatibility between the components of that product. Therefore, technical structure of a system will reflect the operational boundaries of the organizations. This law is applied primarily in the field of software architecture, though Conway directed it more broadly and its assumptions and conclusions apply to most technical fields.

Typically, a Cloud SW project team has a composition that divides the responsibilities to cover every possible facet. People are given functional roles that define their activities. The team definition also includes a *reporting structure,* i.e., there are people whose activities need to be monitored and reviewed periodically to report on their progress. The reporting structure goes up the responsibility chain right up to the top management of the enterprise. Even though each SW project may have a character of its own—it is possible to broadly classify the responsibilities and define roles for people who constitute the project team. Here are some thoughts on the constitution of SW project teams:

1. **Champion in top management:** This person is usually the part of top management. He individually champions the cause of the project. He shows deep involvement in promoting the project for the business value it brings. He also may be a person who maintains liaison with the sponsor of the project or the management team of the client. He keeps contact with the project manager but does not administer the project on a day-to-day basis.
2. **Project manager**: The project manager is the project team leader. He has a direct responsibility towards the progress of the project and meeting all the milestones within the assigned period and the stipulated cost. He is also the single point of contact for the top management executing the project. He also interacts with the client or the people who sponsored the project. The person may be at the level of a Director or Vice President in the organization. He monitors the project regularly.
3. **Software architect**: SW architect is responsible for the design and configuration-related decisions. In fact, the role requires them to define the "*scaffolding*" that holds the structure of the overall system. Structural components at system architecture level are chosen keeping in mind the overall behavior and functional requirements in addition to the non-functional requirements like performance, reliability, etc. He reports to the project manager. In large organizations, SW architects are sometimes loaned from a central pool of experts. Many

enterprises have positions of internal consultants. The internal consultants are assigned to help in chartering the course of projects.

4. **DevOps Engineer:** A Development and Operations (DevOps in short) engineer is an IT expert. This person knows how to support a software project in a private, public, or hybrid Cloud [1]. DevOps engineer's tasks and responsibilities include scripting, QA and metrics collection to help the project management and deployment. This person needs to take care of IaaS, PaaS, and SaaS deployment requirements in the Cloud.
5. **Analysts and subject matter experts (SMEs):** Usually the SMEs are domain experts who can analyze domain-based requirements. They advise the designers who ultimately design the system-based solutions keeping in mind the architecture and domain requirements. In most cases, SMEs operate as internal consultants to the project. Sometimes, this person may come from a customer domain. For example, for a bank-related project, one needs an expert banker to define, understand, or explain banking operations and mandated workflows. The SW architects or designers take the help of SMEs in arriving at their decisions and choices they make.
6. **Software designers:** There may be one or more SW system designers in a project team. In small companies, the designer and SW architect may be the same person. Regardless of how the designers come to be part of the team, they are responsible for the choice of all high-level system component designs—including the choice of operating systems and Cloud environments, interoperability for different virtual machines, databases, programming languages, etc. In very large system designs, they are also the people who decide on the nature of the middle ware to be utilized or locally generated.
7. **User interface expert:** In many organizations, there are user interface designers. Nowadays, UI is a super-specialty in some sense. They may come from graphics domain or may have flexible web system design and services background. With usability becoming a key consideration, more and more organizations have been employing UI experts. They may report to the designers or system architects. In particular, they may design the interface of payment gateways and app's UI for different class of users.
8. **Programmers:** This is a group of people who do the actual coding and report to the designers. They may be engaged in realizing sub-projects within a project with a lead member, often called "team-lead," who reports to the designer.
9. **Test teams:** The test team's task is to verify and validate correctness of the code. They also have a defect control management system. Many organizations provide a test team support from a central pool. Larger enterprises like Microsoft, Google, IBM, Wipro, Infosys, etc. have a test and defect management vertical. They provide support to all the projects in the enterprise. This group takes up the code and tests if it meets the functional and non-functional requirements. The non-functional requirements usually concern with performance, reliability, operations under different load conditions or local regulations, etc.

10. **Installation, configuration management, and maintenance personnel:** Most large corporations maintain a separate group of people for system configuration management (SCM). They provide complete installation support. Note this important role arises from the fact that it concerns security and access rights management. However, from a SW project point of view, installation at customer site is critical. In addition, configuration may need tuning to achieve desired performance levels within the customer's operating environment. System configuration management is an art, and one needs to have good understanding of system's environment. Besides installation, the team must provide for maintenance support and training.
11. **Reporters, document generators, and scribes:** Reporters and scribes may be chosen from programmer teams. They prepare reports from time to time. They also keep the record of progress and report to the project manager. In addition to this, there are two other levels at which documentation needs to be maintained. Some organizations in fact employ document writers to just do that. The two levels at which documents required are:

    (a) **Design documentation:** This is a major software asset that gets created as project evolves.
    (b) **User level documentation:** This is the documentation which is prepared to be handed over to the end user.

12. **External auditors:** The project may need audit from an external auditor to raise the level of customer's confidence in what is delivered. The auditors, while external to the project, should enjoy a certain level of respect and must display neutrality in their task. Auditors may also be from certification agencies. Audit always should result in documenting key findings about the project that get recorded for later use.

Given the inclusion of hardware product engineering into IoT projects, the software engineering roles and responsibilities expand beyond a Cloud software project, while simultaneously satisfying additional constraints. Primarily, an IoT project may involve embedded software design in addition to Cloud software design. The introduction of embedded software requires a different skill set of software engineering and the necessity to collaborate with hardware engineers, possibly in an electronics laboratory environment. Here are additional roles and responsibilities when an IoT project is considered:

13. **Systems Engineer/Product Architect:** A systems engineer/product architect is responsible for ensuring that all aspects of an IoT project are engineered as a single, cohesive system. They ensure that the software design works well with the hardware design and the user interface design of an IoT product. A key aspect of the systems engineer/product architect role is to define the requirements and interfaces between all engineering components of the project.
14. **Hardware Engineer:** One or more hardware engineers may be necessary to design and develop the electrical and mechanical aspects of an IoT product. Most often, hardware engineers have experience in electrical or mechanical

engineering. They are responsible for capturing schematics, layout of printed circuit boards, and design of physical enclosures of the electronics.

15. **Embedded Software Engineers/Programmers:** This is a group of people who do the actual coding for embedded electronics and report to the designers. The difference between Cloud and embedded software engineers is in their skill sets. Embedded software engineers and programmers need to be familiar with low level software languages such as C and C++ and may also need to know hardware description languages (HDL) such as VHDL or Verilog. Embedded software engineers need to be cognizant of limited memory and compute power available on IoT devices.
16. **Human Factors Engineer:** Human factors engineer can be thought of as the user interface experts described above, but whose responsibility includes both software and hardware. For IoT products, it is important that a user can interact with the device in a safe and effective way. Human factors engineer ensures that a user can understand all aspects of the device as well as its software interface.
17. **Manufacturing Engineer:** An IoT device must have a physical embodiment—a device that can be packaged, shipped, unboxed, and used in a household, commercial, or industrial environment. A manufacturing engineer or team of manufacturing engineers is necessary to ensure that an IoT device can be built repeatably, reliably and in a manner to withstand all the external stresses it is expected to experience throughout its lifetime.

The additional roles and responsibilities typical for an IoT project are not mutually exclusive from the software project roles. In fact, synergies often form between the software and IoT project team members. A noteworthy example is the DevOps engineer. When included in an IoT project, the DevOps engineer's efforts supporting continuous integration and continuous delivery (CICD) take on added meaning with hardware-in-the-loop (HIL) testing. User interface engineers collaborate closely with human factors engineers and project managers manage the entire software and hardware IoT project.

The above enumerated roles and responsibilities are only representative. Projects are populated in accordance with the needs befitting the size of project. Large enterprises can afford and will usually have larger teams because they mostly work on big projects. For instance, companies like Infosys have internal teams that take up turnkey computerization projects to create IT infrastructure to support their FINNACLE banking solution. Also, IBM has teams that implement automation to support Cloud-based billing and flexible tariff on offer for the utility companies.

A fairly large number of projects actually get done by smaller teams. In smaller teams, people take up multiple roles. For example, one of the co-authors was the champion, architect, and designer in his own organization. There have been times when our first co-author has also been a test manager. In some organizations, programmers are identified to play the role of testers for each other's code. Programmers also may prepare design and test documentation. There are many programmers who have a flair for system activity and they double up as people who do the installation and configuration management.

However, there are some roles that are distinct—that of the analyst and also there can be only champion at the top management level. Later, when we discuss SW project outsourcing and new business models like software as a service (SaaS), we can identify other distinct roles.

Regardless of how the roles and responsibilities get assigned, the *success of a project largely depends upon the way people with assigned roles take the ownership* of the tasks in their sphere of activity.

1.8 The Rest of the Book

In earlier sections, we have defined software project and its life cycle. In this process, we also recognized that many of the concepts of PMBOK apply to Cloud-based SW projects as well. In addition, the third phase (ref. Fig. 1.3), i.e., the SW development has its own life cycle and there is much to study in respect of the models used there. We also noticed that in SW projects, the activities of people are critical to its success. The project should have a strong quality assurance and test regime. Keeping these points in mind, we have organized the rest of the book as follows.

Chapter 2 introduces the IoT-based devices and systems. It explains basic differences between IoT and Cloud, then delves into the hardware, software, and networks to support IoT deployments. It ends with a few cases studies of IoT usages and management.

Chapter 3 addresses Cloud and SW project business cases, starting with an introduction of how to make a good business case. Then it dives into the intricacies of information technology project management (ITPM) details, listing the roles and responsibilities of various actors. The appendices at the end of this chapter provide a business plan template, a project charter template, and some basics of PERT and critical chain scheduling. Following this, in Chap. 4, we explore Cloud-based product development process, starting with traditional software development methods. Here we emphasize the need for and importance of gathering engineering requirements, undertaking Agile development, performing DevOps, and Cloud-based product testing. We end the chapter by looking at various ways to manage a product roadmap, including change control methods and partner management, etc.

In Chap. 5, we delve into IoT-based application development processes, starting with an understanding of sensors, prototyping hardware, embedded software, and need for low-power operations. We conclude this chapter with connecting IoT to Cloud and challenges in IoT testing and related project management.

In Chap. 6, we describe planning strategies for Cloud-based projects, resources requirements, project execution frameworks, managing besides how to maintain budgets, service level agreements, etc. We end the chapter by reviewing roles and responsibilities, and ways to monitor Cloud-based projects. We reckon that a project's success also depends on the communications between the stakeholders. We discuss this aspect in Chap. 7. We examine the limits of technology and using human elements to overcome those. Our description of communication includes the process of reviewing and managing a project across time zones with remote teams contributing to the success of a Cloud-based project. The chapter concludes with a sample of regulatory requirements.

Chapter 8 covers IoT-based deployments, maintenance, updates, and management issues in the field.

At this stage, it was felt necessary to introduce a major application to demonstrate how a Cloud-based medical solution was conceptualized and realized. Note that the end users in this case are the hospitals with prescribed internal organizational practices, strong local and federal regulations. Chapter 9 is entirely devoted to showcase how a Cloud-based product was delivered in a regulatory environment, while overcoming IT and data management challenges.

Chapter 10 deals with a new class of hybrid Cloud-based applications, covering their performance and security aspects. We look at some real-life hybrid applications and how organizations need to evolve and manage to develop these.

Having described one case study in detail, we felt it necessary to describe and summarize the lessons learned from the previous chapters in Chap. 11. This takes the form of some rules of thumb, including pitfalls to avoid and to focus on the important things for success. We revisit Agile processes and examine some additional challenges in migrating an existing product to Cloud. As we approach the end of this book, we explore new horizons for Cloud-based opportunities for value creation in Chap. 12. We look at three examples of Salesforce, Microsoft, and Netflix as companies that either started a new Cloud-based project or migrated their existing products to different business models to leverage the Cloud and create new values for their respective businesses.

The goal of this book is to extend and apply principles of software engineering to harness IoT and Cloud-based products and service opportunities.

Points to Ponder

1. What are the key considerations in deciding if a software project ought to be deployed in a Cloud vs. onsite infrastructure?
2. How does a software project's development life cycle differ between following a Waterfall process vs. Agile methods?
3. Why do the success rates vary significantly between medium or large projects following Waterfall vs. Agile development methods?
4. How does an error in a SW differ from a defect in SW?
5. What do the vertices represent in a project's cost triangle?
6. If an organization's product evolves from a legacy application to be hosted in the Cloud, how may its organization structure change to facilitate this evolution?
7. What are the "Things" in the term ""Internet of Things" (IoT)?

Answers

1. What are the key considerations in deciding if a software project ought to be deployed in a Cloud vs. onsite infrastructure?

(a) Key considerations for a project's deployment in a public vs. local infrastructure are the IT infrastructure ownership and maintenance responsibilities.
(b) For onsite enterprise software deployment, the home organization is responsible for maintaining all hardware and software infrastructure. The organization is in full control of solving all related issues.
(c) For public Cloud deployment, the project team will be dependent on a third-party provider. So, application's performance, security, and deployment of software need to follow the practices prescribed by the Cloud vendor. The advantage is that it relieves the project organization from the responsibility of undertaking frequent infrastructure updates.

2. How does a software project's development life cycle differ between following a Waterfall process vs. Agile methods?

(a) Waterfall is a sequential methodology and tasks are handled as a linear process. The specifications are fixed at the start of the project.
(b) Agile is an iterative process that allows flexibility in a product's requirement changes.

3. Why do the success rates vary significantly between medium or large projects following Waterfall vs. Agile development methods?

(a) Agile is more reflective of how the real world works as requirements often change with the evolving user needs, competitive product offerings, and a project's own execution outcomes. A Waterfall process is generally not able to absorb the changes in requirements. This results in more failures, unless the project is small and quick to implement.

4. How does an error in a SW differ from a defect in SW?

(a) The SW errors may creep at any stage. If detected early, it is at best an error. If an error manifests later in the SW development, then it is marked as a defect.
(b) If an error is caught at an early stage, before the architecture, design, or coding are frozen as it is easier to alter any of these elements. An upstream error that escapes manifests as a defect downstream.
(c) Making changes later in a project's life cycle is harder, disruptive, expensive, and sometimes even be infeasible. This may entail both upstream and downstream changes resulting in a lot of work to be undone or redone.

5. What do the vertices represent in a project's cost triangle?

(a) Features: Features outline what a product is expected to deliver.
(b) Resources: Resources refer to human capital, computer resources, and other inputs consumed.
(c) Timelines: It represents time allocated to implement and deliver a product from start to end.

6. If an organization's product evolves from a legacy application to be hosted in the Cloud, how may its organization structure change to facilitate this evolution?
   (a) As we learnt in Conway's law [13], an organization can best design a product that mimics its structure. So, if a legacy to evolve for hosting in a Cloud, then the organization may need a separate team with expertise in the Cloud. It also needs a cross-team architect to decide on how to best partition the legacy and design the APIs between the native and hosted parts of the application.
   (b) This is further covered in Chap. 10.
7. The "things" in IoT are physical devices with electronics that run embedded software and typically include:
   (a) Sensors to measure the real-world entities in the environment of interest.
   (b) Actuators to interact with the real-world environment.
   (c) A connection to a local network or the Internet, to monitor, or regulate specific parameters.
   (d) A software or Cloud interface application.

References

1. Sehgal, N. K., Bhatt, P. C. P., & Acken, J. M. (2020). *Cloud computing with security*. Springer. https://link.springer.com/book/10.1007/978-3-030-24612-9
2. Oliveira, F., Costa, D. G., Assis, F., & Silva, I. (2024). Internet of Intelligent Things: A convergence of embedded systems, edge computing and machine learning. *Internet of Things, 26*, 101153. ISSN 2542-6605. https://doi.org/10.1016/j.iot.2024.101153
3. https://netraclos.com/what-are-iot-devices/
4. Project Management Institute. (2021). *A guide to the project management body of knowledge (PMBOK)*. PMI Publishing, PMI. https://www.pmi.org/pmbok-guide-standards/foundational/PMBOK
5. Royce, W. W. *Managing the development of large software systems*. IEEE WESCON (1970); Also presented at 9th International conference of Software engineering.
6. Marchekwa, J. T. *Information technology project management*. Wiley 1st edition 2003, 3rd edition 2009, Also see course slides by Prof. Marchekwa, North Illinois University.
7. Standish Group. (2015). https://www.standishgroup.com/sample_research_files/CHAOSReport2015-Final.pdf
8. Bhatt, P. C. P. (2021). *Software design, architecture and engineering: Concepts and practice*. Prentice Hall India, Easter Economy Edition. ISBN-978-81-951611-4-0.
9. http://www.agile-software-development.com/2009/09/agile-project-management-software.html
10. Frederic Brooks. (1979). *The mythical man month*. Addison Wesley.
11. Cognitive dissonance. http://en.wikipedia.org/wiki/Cognitive_dissonance
12. McConnell, S. (1996). *Rapid development*. Microsoft Press.
13. McConnell, S. (1997). Upstream decisions, downstream costs. *Windows Technical Journal*. Article available on Web: http://www.stevemcconnell.com/articles/art08.htm
14. Larman, C. (2007). *Agile and iterative development: A manager's guide*. Pearson Education.
15. https://en.wikipedia.org/wiki/Conway%27s_law

Chapter 2
Introduction to IoT Products and Systems

2.1 Introduction

IoT-based devices and systems are growing in popularity, adoption, and scope. IoT analytics has forecast 14% growth of the global market for IoT products, resulting in over 40 billion connected devices by 2030 [1] (Fig. 2.1).

The services rendered by an IoT device or system can include personal data tracking up to core infrastructure monitoring including advanced agriculture, city utility, and service planning. Some applications have little cybersecurity or safety implications while enabling critical functions in various industries. It is helpful to categorize IoT products into four main groups [2]:

- **Industrial IoT (IIoT)**
  - The Internet of Things as applied to industrial settings with a goal to improve efficiency or logistics. Radio frequency identification (RFID) systems are used in a variety of applications, including supply chain and inventory tracking, process flow monitoring, water quality monitoring, and energy management systems.
- **Commercial IoT**
  - This category includes many products and can be subdivided into consumer and enterprise IoT. Personal fitness trackers and home security systems are examples of consumer IoT products. Enterprise IoT focuses on business solutions including improving quality of service. The healthcare industry is a good example. Remote patient monitoring, pharmaceutical tracking, and robotic surgery IoT products are improving the quality-of-care patients receiving every day.

P. C. P. Bhatt et al., *Project Management in Cloud and IoT Applications*,
https://doi.org/10.1007/978-3-031-99427-2_2

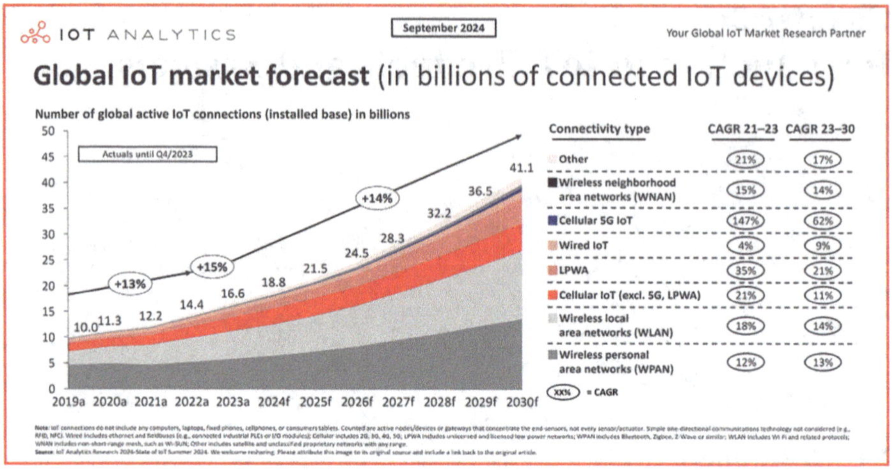

Fig. 2.1 Global IoT market forecast [1]

- **Infrastructure IoT**
  - The amount and type of data that can be provided by IoT products is enabling rural and urban planning to achieve new levels of efficiency and service. Traffic management, air quality monitoring, and predictive city service needs such as rubbish removal are examples of the benefits of Infrastructure IoT.
- **Defense IoT**
  - Countries have developed IoT products to address their specific military needs including enemy movement monitoring and operational readiness. Surveillance drones, smart combat helmets, and soldier health monitors are just a few examples of devices that can support military operations.

All applications can be distilled to the fundamentals of an IoT application—interactive hardware with embedded software and the potential to connect to Cloud-based applications. This chapter delves into those fundamentals and concludes with some detailed IoT use cases.

2.2 IoT and Cloud

This book details the management of both Cloud-based software applications and IoT applications. It is important to understand the boundaries between Cloud-based software and IoT systems. To do this, an understanding of the key components of an IoT system is beneficial. Figure 2.2 shows these key components.

Starting from the left in Fig. 2.2, we see that IoT systems have a hardware platform which includes embedded software and local data storage. The temperature regulating system in a smart home is an example of IoT hardware. The embedded software in

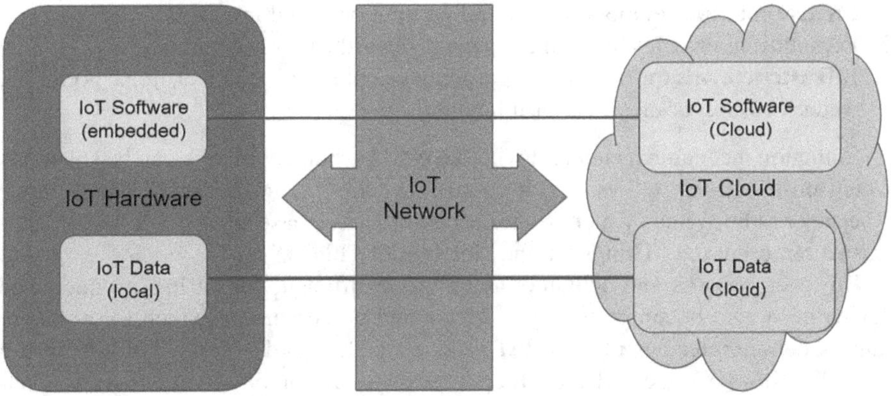

Fig. 2.2 Key components of IoT systems

IoT products is resource constrained and must be designed to run on microprocessors or system-on-chip. This is unlike Cloud software which may be resource rich and running on multiple high-power servers. Similarly, the data which may ultimately be transmitted to the Cloud and stored in large quantities for an IoT system will need to momentarily reside on the IoT hardware's limited on-board memory.

In the middle of Fig. 2.2, the IoT network is shown. This represents both the communication technology, and the protocols used to connect IoT software and data which reside on the IoT hardware to the IoT Cloud. The smart home temperature regulating system mentioned above could connect to the Internet via Wi-Fi. In this case, Wi-Fi would be considered the IoT networking technology, whereas the data from the temperature regulating system may be transferred using Transmission Control Protocol (TCP). The IoT network can be thought of as the boundary between IoT systems and the Cloud.

Apart from the IoT Network, a complete connected IoT system will include a portion in the Cloud, shown on the right of Fig. 2.2. Some IoT systems may only use the Cloud for data storage—in which case the IoT Cloud software portion will be small. However, other systems like the smart home temperature regulating system may include a full-featured application with user interface that enables certain controls, like setting room temperature, to be performed from remote locations. This chapter primarily introduces IoT for the required minimal understanding. Later chapters, Chaps. 5 through 9, discuss the Cloud support and other hardware, software, and networking in greater detail.

2.3 The Hardware and Software of IoT

The hardware in IoT products typically includes sensing mechanisms and/or actuation mechanisms. Sensing mechanisms measure aspects of real-world parameters. We briefly mention the three categories:

1. **Environmental:** temperature, humidity, light or sound intensity.
2. **Personal**: biometrics including a person's heartbeat or breathing.
3. **Infrastructural:** traffic density, particulate content (air quality), noise pollution, water pressure, or electrical grid loading.

Actuation mechanisms interact with the world around us. These can be solenoids which push buttons, relays which control switches, electric motors which move objects, or radio signals which trigger behavior from a base station. All these mechanisms represent the "Things" in the "Internet of Things."

However, sensors and actuators alone are insufficient for an IoT product. The device must also be supported by software and processors on which the software runs. A common term applied to IoT products is the word "smart." For example, a smart lighting system can detect when it needs to turn on or off. Smart waste management systems can alert city officials when trash bins are empty, full or ready to be picked up. The "smart" aspect of IoT products is enabled by IoT software. Digital signal processing of sensor data and software control of actuators may be needed. In some IoT devices, the processing is done entirely by the electronics hardware contained inside the IoT device. At the very minimum, embedded software is necessary, but in most IoT devices, Cloud software is recommended for an in-depth and predictive analysis or "smart operations" [3].

One way to visualize the hardware and software of IoT is to consider a common closed-loop control system, as shown in Fig. 2.3.

In such a system, a controller collects and compares inputs and feedback to drive a process with the goal of achieving a desired outcome. Imagine an Internet-connected smart agriculture irrigation system that optimizes irrigating the crops. Input is provided to the smart irrigation system in the form of a set point—the desired moisture level for a particular crop. Input is also provided through feedback from remote sensing of moisture levels with ground water moisture sensors. With this input, the hardware controller of the smart irrigation system can pre-process the moisture data with on-board embedded software. Additional analysis which may be needed can be performed in the Cloud to complete the control loop processing. For example, the control loop could account for rainy day conditions versus drought conditions. The processing could also identify outliers in sensor readings due to animal or other interference. Once the input and feedback data has been processed, the irrigation system can activate switches to water sprinkler systems to match the desired water moisture level. Figure 2.4 shows the smart irrigation system as a control loop with IoT hardware and software identified.

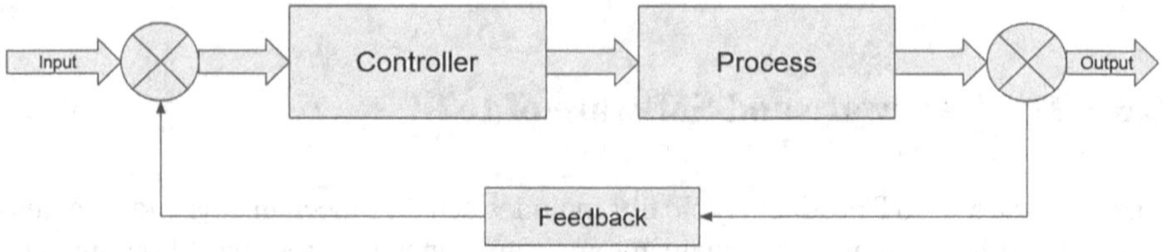

Fig. 2.3 Schematic of closed-loop control system

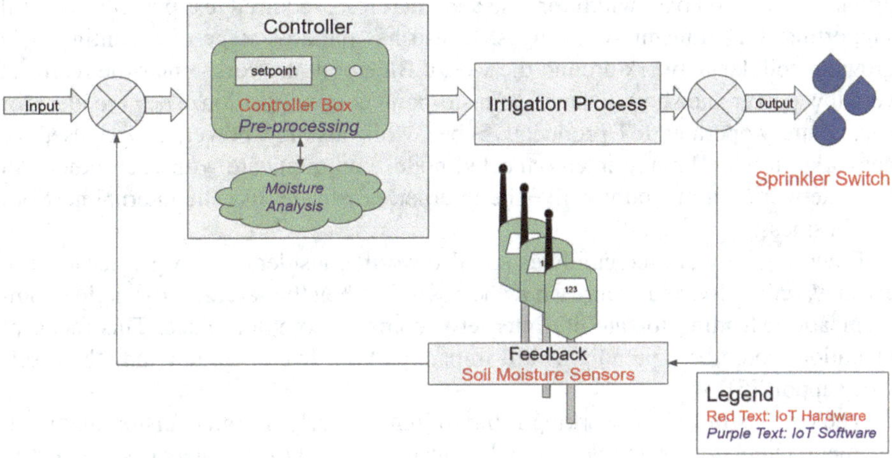

Fig. 2.4 Smart irrigation IoT hardware and software as control loop

Many IoT products work in a manner like the smart irrigation system described. A combination of sensors, electronics to communicate and acquire sensor data, embedded and Cloud software to process the data, and actuators to interact with the real-world environment is typically needed in an IoT system. The above example also alludes to another crucial component of IoT systems: the IoT Network.

2.4 The Networks Behind IoT

When dealing with the Internet of Things, the term "network" has multiple interpretations. An IoT network can refer to a connected group of physical things like a set of smart light bulbs in a house. It can also refer to the mode of communication between IoT products and the Cloud. It may also sometimes refer to a specific communication protocol used by a networked device, such as Ethernet. In general, one needs to consider both the networking technology and protocols used to describe IoT networks. IoT systems can have local, ad-hoc networks (no Internet connectivity), or connected networks (with Internet connectivity). Examples of local networks include Bluetooth and NFC. The examples of connected networks include Wi-Fi or satellite modem. For Internet-connected devices, a corresponding Cloud application is necessary to take advantage of the IoT product. In fact, a Cloud application can also enhance and add functionality to the entire product above and beyond what is possible with an off-line network.

Common IoT networks include Wi-Fi, Cellular, Bluetooth, and Zigbee [4]. Wi-Fi networks are commonplace, easy to access and provide fairly reliable data transmission over ranges sufficient for homes or businesses. However, low-power IoT products may choose not to use Wi-Fi given its power requirements. Cellular IoT provides a reduced bandwidth for data and, therefore, required less power while still supporting long transmission ranges. It also has the advantage of a continuously growing cellular network around the world. Bluetooth networks enter the realm of very low power and short distance point-to-point communication. They are ideal for small battery operated IoT products. Zigbee, while also low power, is often used for networks in which many interconnected nodes communicate with each other. All these networks can be public, like the Internet, or private, like the internal network of a business.

Emerging network technologies are also worth consideration. A greater level of efficacy, reliability, and security may be possible when the satellite-to-satellite communications leading to satellite Internet becomes a common place. This mode of operation would be a paradigm shift impacting both IoT operation and Cloud service support [5].

The options in IoT networking bring to light an early design decision that management and engineering teams should consider. The choice of network connectivity hardware has repercussions throughout the entire product and project. Note that networking with a wireless technology, antenna and power considerations can be expected to constrain other design choices such as product size and usability. Radio frequencies used for communication may be available for unlicensed public use or may need government approval for transmission adding regulatory and test burden, as described in Chap. 9. It is a good practice for the product teams to carefully understand their use scenarios, operational environment, and data communication needs to make an informed decision on choice of IoT network.

2.5 The Data in IoT

IoT products have a special ability to collect data from the real world. As opposed to a software product, the hardware in IoT electronics typically includes one or more sensors that can measure environmental data in real time. This data is valuable for both consumers and developers of IoT products. To maximize the utility of data in IoT systems, it is useful to consider how the data is managed end-to-end, i.e., from the time of data collection to where the insights are provided to consumers.

Sensors measure analog aspects of the environment such as temperature, pressure, motion, or light. An IoT product uses Analog-to-Digital converters (ADCs) to digitize these into data for further processing. In most cases, the data is processed using advanced algorithms which may rely on prior measurements. In addition, the data collected may need to be modified (clipped, converted to different data types, or decimated) prior to being processed. This is referred to as pre-processing. Local data storage on an IoT product enables on-board pre-processing and allows for some level of more advanced processing before data is transmitted to the Cloud for further insights. Based on the processing bandwidth of the IoT hardware, an appropriate amount of low-level memory devices such as RAM or ROM should be available on the IoT product.

Consider a smart home lock device with a touch screen for a four-digit access code entry. This device will digitize the pressure points used when a person enters a code for access. The smart lock will convert the touch locations into numbers based on the lock screen shown. For security, the home entry code is not stored on the physical lock device but is stored in the Cloud. As such, the device will collect and store all four digits entered so that an entire four-digit access code can be transmitted to the Cloud for verification. Once verified, a command can be sent back to the smart lock to allow access.

The smart lock example illustrates that bi-directional data transmission should also be considered in IoT products. Data transmission is enabled by the IoT network chosen. For applications like the smart lock, there is a low data bandwidth requirement—not much data need be transmitted back and forth for the lock to operate. Other IoT products, such as security cameras where high resolution video data is used, may have a higher data bandwidth requirement. It is important for IoT project management to consider data management throughout the design process. An IoT project must not only manage data on the physical device, but also how that data can be transferred and stored on the Cloud.

2.6 Symbiotic Relationship Between IoT and Cloud

A synergy is formed when IoT is coupled with Cloud services. The infrastructure and resources available in the Cloud help in gainfully enhancing the utilization of processing power available in self-contained electronics devices. The hardware present on an IoT product provides the Cloud applications with abilities to sense, evaluate, and interact with the real world. Together, they benefit both consumers and businesses that use or build them. The enhanced data storage and processing proffers scalability, flexibility and provides deep operational insights [6]. In brief, when IoT and Cloud applications work in tandem and in harmony, the horizons of applications get extended.

Every commercial IoT product captures a great amount of data in its day-to-day use. When considering an entire fleet of commercial IoT products, the total data managed may scale more than linearly. For example, smart watches may collect personal fitness data when used in isolation but could also collect data on nearby friends' activities if used in a mode allowing shared data. Combined with the massive data requirements for industrial, infrastructure, and defense IoT, the overall storage needed is enormous. One forecast from the International Data Corporation (IDC) estimated nearly 80 Zettabytes (one ZB is 10^{21} bytes) of data will be generated by IoT products in 2025 [7]. Therefore, the benefits of Cloud storage for IoT become evident. The Cloud offers a platform for potential analytics with a massive data storage to process IoT product's data in failsafe mode of operation on remote servers.

The Cloud's flexible and scalable design is of great benefit to the providers of IoT products. Rather than continuing to invest in additional hardware to manage and process IoT data as the product increases in its market share as new and innovative usages emerge. The Cloud offers dynamic resource allocation which can respond to business needs for shifting product demands [8]. This flexibility also allows for minimizing overhead during short- or long-term downturns in product usage, providing opportunities to maximize efficiency and reduce the overall cost of a product.

Data insights refer to meaningful information culled out of datasets through analysis to make informed decisions or better understand a particular subject [9]. The processing power, data storage capabilities, and scalability mentioned earlier all enable the Cloud to provide valuable data insights that simply cannot possibly be done in IoT hardware alone. For example, traffic monitoring IoT products can be placed strategically throughout a metropolitan area to upload data to the Cloud. An AI algorithm deployed in the Cloud can mine this data and provide predictive traffic patterns as an insight for city planners. This would help to evaluate and strategize to reduce congestion. When the Cloud is integrated with IoT, insights like this can add value to every IoT product across any industry.

2.7 IoT Case Studies

A short summary is provided below for a few cases where companies or organizations benefited by the adoption of IoT products. These case studies exemplify how IoT and the Cloud can be used to meet the needs of businesses and consumers. References are provided for each case with additional details. Many more examples like these exist and can be found on the Internet.

2.7.1 Xylem [10–12]

Sector Utilities and smart cities.

Application Infrastructure for sustainable water use.

Background Xylem is a global water technology provider who strives to empower their customers to build a more water-secure world. Over 10 years, they have grown from a pump manufacturer to a technology and service provider with a diverse portfolio of water and wastewater treatment solutions. As of March 2025, the company serves customers in more than 150 countries, achieving $8.6B in revenues.

Challenge Rapid population growth in areas of water scarcity combined with aging water infrastructure pose a growing problem for effective and safe water management. Water utility managers have to deal with pipe leaks, main breaks, sewage and storm water overflows, and other serious issues. This, coupled with the lack of adequate funding in water infrastructure, demands an efficient solution for sustainable water management.

Results Xylem introduced Avensor, a Cloud-based IoT solution for smart monitoring and remote control of water assets. Avensor collects and analyzes data on pumps, mixers, UV units, and more. It provides users with live data, trends, and real-time alerts. Avensor has been installed at more than 2000 stations, for 700+ customers in 30 countries. With Avensor, customers that leverage the IoT-enabled data have been able to improve first-time fixes by 11%, increase asset uptime by 9%, and reduce mean time to repair by 9%.

2.7.2 MachineMetrics Predictive with BC Machining [13, 14]

Sector Manufacturing.

Application Predictive maintenance.

Background BC Machining was founded in 2007 and specializes in machining fabricated metal products. They contract with medium-scale businesses and Fortune 500 companies offering products for various industries including healthcare, defense, transportation, and power tool manufacturing.

Challenge As a machine shop, every ounce of scrap produced during operation is a cost to the company. Their Swiss-type CNC-operated lathe had an issue where they would break tool bits but continue to run because they were operating with an automated process. This issue impacted the quality of output and required an operator to walk over to the machine to verify whether the tool was broken or not.

Results MachineMetrics, an intelligent platform for manufacturing success, was installed. This IoT product combines sensors which capture data from machines, operations, and shop floor systems. With their Connectivity Hub, data across various makes and models of manufacturing equipment is standardized and connected to the Cloud. MachineMetrics gave BC Machining the ability to track all scrap produced and the operator involved. With this data, BC Machining can optimize process and training to reduce scrap. MachineMetrics also provided 100% detection of broken tools in the Swiss turn machines and predicted tool breakage to eliminate all scrap from automated machines. Operators at BC Machining can now focus on producing parts, rather than checking on machines to see if they are producing defects.

2.7.3 *IoT for Interactive and Collaborative Learning [15]*

Sector Education.

Application Interactive and collaborative learning in higher education.

Background The transformative nature of IoT technologies has impacted many industries, including education. IoT products can provide hands on experience in real-world data monitoring and management to students. The use of learning analytics with IoT is an active area of education research.

Challenge With recent global developments such as COVID and regional conflicts, environmental awareness is of particular interest to students in higher education. IoT is well suited to environmental monitoring and can provide valuable insights. However, for students to fully take advantage of IoT in learning, it is necessary to understand their digital data competence and the effectiveness of applied IoT projects in course studies.

Results Two case studies were performed to explore the integration of environmental awareness and IoT technology. The first case study focused on two key aspects of an IoT-based solution: students' comprehension of the international units used for environmental data (IoT inputs) and current recommended levels of carbon dioxide and indoor ventilation (outputs which can be controlled via IoT products). The results of the first case study showed the gaps in students' knowledge regarding environmental data and informed the second-case study. The objective of the second-case study was to challenge students to create a game leveraging data from an IoT system that captured environmental variables. Results from this study show that students were able to utilize the data collected by the IoT sensor system to interpret and understand regulations related to ventilation and energy conservation. The games created promoted practical learning and knowledge application. Overall, these results highlight the importance of integrating IoT technology into higher education. IoT technology offers opportunities for real-time data collection, personalized feedback, and the creation of more meaningful learning experiences.

The three case studies cited above were chosen from different domains to demonstrate the nature of operations where IoT could be gainfully employed. In later chapters, we shall deal with many other aspects related to IoT hardware, software, networking, and Cloud support.

2.8 Summary

This introductory chapter focuses on Internet of Things (IoT) products and their relation to the Cloud. We explore a broad spectrum of IoT industries to identify basic building blocks of IoT devices and systems. The hardware, networks, and data associated with IoT are spotlighted throughout the chapter. The benefits of incorporating Cloud applications with IoT products are also explained. The chapter concludes with IoT case studies taken from real-world applications to showcase the potential that the IoT and Cloud have for delivering high value propositions in the form of IoT products.

Points to Ponder

1. How does Industrial IoT differ from Infrastructure IoT?
2. Explain how an IoT product can relate to a closed-loop control system?
3. List a few networks used in IoT and describe the differences of those networks in respect of their protocol, performance, and context of use.
4. What benefits are obtained when combining an IoT product with a Cloud application?

Answers

1. How does Industrial IoT (IIot) differ from Infrastructure IoT?

   (a) Industrial IoT can be used for monitoring and regulating industrial process flows such as inventory tracking in a supply chain or monitoring key operational conditions in manufacturing. These include production process temperatures, coolant flow, energy management systems, and process throughput. All these applications are geared towards industrial settings with goals in improving efficiency. IIoT uses smart sensors, actuators, and radio frequency identification to achieve these goals.
   (b) Infrastructure IoT includes products that may monitor air or land traffic, air and water quality, or waste disposal. Infrastructure IoT is enabling rural and urban planning to achieve new levels of efficiency and service. The concept of a smart city is one in which infrastructure IoT is heavily integrated into daily life by providing services for the citizens and businesses of the area through digital data collection and Internet connectivity.
   (c) The typical customers for Industrial IoT are businesses in the manufacturing or storage sector while the typical customer for infrastructure IoT are local governments, public or private services, and community organizations.

2. Explain how an IoT product can relate to a closed-loop control system?

   (a) Key aspects of a closed-loop control system are the inputs, controller, process, outputs, and feedback.
   (b) Inputs—IoT products relate to control system inputs through their sensors and data collection. Sensors are used to collect information needed to affect a process and understand how the system's output impacts the control loop goals.
   (c) Controller—In an IoT product, the electronics hardware platform will have a microprocessor, microcontroller, system-on-chip, programmable logic, or a combination of these. This is the part of the system that operates on the input data acquired.

(d) Process—The process of a control system can be thought of as the use case or intended use of an IoT product. For example, an IoT home control system may have a room heating process in which temperatures are monitored and heating or cooling elements used to adjust the temperature to a set point.
(e) Outputs—Outputs of a control system relate to actuators, motors, switches, and other IoT hardware for interacting with the real world. The relationship can be indirect when an IoT product controls other hardware, such as using motors to automatically open windows for increased air flow, or direct, such as using an integrated heating element to increase temperature.
(f) Feedback—Feedback and inputs are similar in nature. They are both collected with the sensors of an IoT product and fed into the controller to achieve the desired outcome of the control system.

3. List a few networks used in IoT and describe the differences of those networks in respect of their protocol, performance, and context of use.

   (a) Wi-Fi. Wi-Fi networks provide fairly reliable data transmission over ranges sufficient for homes or businesses at the cost of power and protocol complexity. Wi-Fi is part of the IEEE 802 protocol family and requires a somewhat deep communication stack to be implemented on hardware. However, the ubiquity of Wi-Fi in populated areas today ensures that it can be used in a wide variety of contexts.
   (b) Cellular. Cellular IoT requires less power but is also limited in data bandwidth. Cellular networking is now a mature and reliable technology, and cellular coverage includes densely populated areas like cities but also rural areas. While some very remote areas may still not have coverage, access to high speed cellular networks in the form of 5G, 4G, and 3G is far spread. This, in combination with the lower power requirement of cellular compared to Wi-Fi, make cellular networks a good candidate for IoT products in remote or developing areas.
   (c) Bluetooth. Bluetooth networks are low power and simple to use but support shorter distances and low bandwidth comparatively. Bluetooth networks are ideal for IoT devices that need to network locally with other devices or gateways to the Internet. Its low power requirement makes Bluetooth a good candidate for devices that operate on battery power. Zigbee is another network protocol like Bluetooth.

4. What benefits are obtained when combining an IoT product with a Cloud application?

   (a) In summary, the Cloud's capabilities such as data storage, processing, dynamic resource allocation, and data analytics add great value when combined with IoT products. These complement the constrained resources present in IoT hardware and open the design space for products to provide higher value propositions to the end user. Similarly, the real-world data collected by IoT systems provide Cloud applications with an additional dimension to operate as compared to strictly software applications.

References

1. https://iot-analytics.com/number-connected-iot-devices/
2. https://www.spiceworks.com/tech/iot/articles/what-is-internet-of-things/
3. https://www.oracle.com/scm/smart-operations/
4. https://www.particle.io/iot-guides-and-resources/iot-protocols-and-standards/
5. Yurong, H., & Li, V. O. K. (2001). Satellite-based internet: A tutorial. *IEEE Communications Magazine, 39*(3), 154–162. https://doi.org/10.1109/35.910603
6. https://www.hologram.io/blog/cloud-computing-and-iot/
7. https://blogs.idc.com/2021/01/06/future-of-industry-ecosystems-shared-data-and-insights/
8. https://cyberpanel.net/blog/iot-and-cloud-computing
9. https://www.zuar.com/blog/what-are-data-insights/
10. https://iot.telenor.com/iot-case/xylem/
11. https://www.xylem.com/en-us/
12. https://www.xylem.com/siteassets/brand/xylem/resources/brochure/xylem_avensor_2022_brochure_low_en-us.pdf
13. https://www.machinemetrics.com/platform
14. https://www.machinemetrics.com/machinemetrics-predictive-interview-case-study
15. Tabuenca, B., et al. (2024). Unlocking the potential of IoT for interactive and collaborative learning: Case studies in higher education. In M. E. Auer, U. R. Cukierman, E. Vendrell Vidal, & E. Tovar Caro (Eds.), *Towards a hybrid, flexible and socially engaged higher education. ICL 2023* (Lecture Notes in Networks and Systems) (Vol. 901). Springer. https://doi.org/10.1007/978-3-031-53022-7_43

Chapter 3
Cloud and SW Project Business Cases

3.1 Introduction

Cloud computing [1] is an umbrella term and broadly refers to provisioning a wide range of hardware and software IT services to support applications, and data using dynamically scalable pool(s) of remote servers transparently located somewhere on the Internet infrastructure. Such an arrangement accrues the advantage that users need not be concerned about the physical location of the server or storage. According to NIST, the definition of Cloud computing is still evolving [2], to cover all the implied notions and scope. Their current definition for Cloud computing is "a model for enabling convenient, on-demand network access to a shared pool of configurable computing resources (e.g., networks, servers, storage, applications, and services) that can be rapidly provisioned and released with minimal management effort or service provider interaction." Armbrust [3] provides another, similar, definition of Cloud computing as "the applications delivered as services over Internet and the hardware and system in the datacenters that provide these services." The services themselves have long been referred to as Software as a Service (SaaS) [3]. The Cloud, in fact, can also include Infrastructure as a Service (IaaS) and a Platform as a Service (PaaS). These basic service models are defined, starting from the top of a symbolic pyramid, as shown in Fig. 3.1 and described as follows.

1. **Software as a Service (SaaS):** SaaS focuses on the end users of Cloud, to provide them with application and service access so that different users can execute the application binaries in their own virtual machine or server instance(s). These sessions may be running on the same or different underlying hardware, and SaaS enables application providers to upgrade or patch their binaries in a seamless manner. Examples of SaaS providers are Salesforce.com providing CRM (customer relationship management), Google.com serving docs and mail, etc., all of which are hosted in Cloud.

P. C. P. Bhatt et al., *Project Management in Cloud and IoT Applications*,
https://doi.org/10.1007/978-3-031-99427-2_3

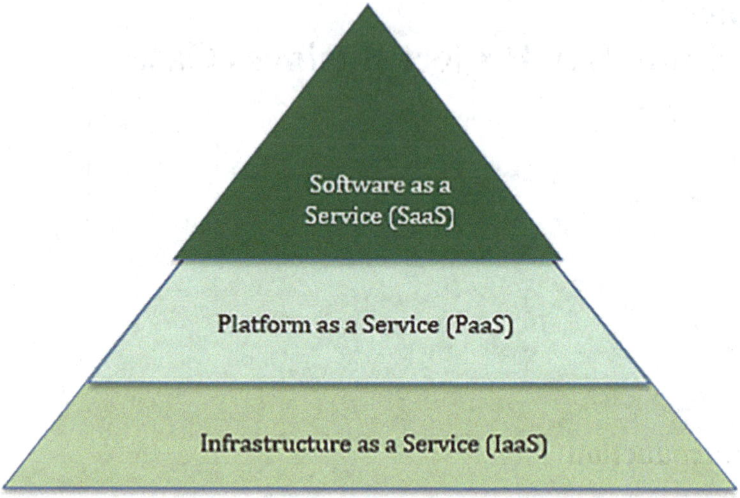

Fig. 3.1 A layered representation of Cloud-based services [1]

2. **Platform as a Service (PaaS):** PaaS focuses on application developers with varying computing needs as required at different stages of their projects. These are met by servers that can vary in number of CPU cores, memory, and storage as required. Such servers are called elastic servers. Their services can auto scale, i.e., new virtual machines can be started for load balancing with a minimal administrative overhead. Examples of PaaS providers are Google's App Engine, Microsoft's Azure, Red Hat's Makara, AWS (Amazon Web Services) Elastic Beanstalk, AWS Cloud Formation, etc. These Cloud service providers (CSPs) have the capability to support different operating systems on the same physical server.
3. **Infrastructure as a Service (IaaS):** IaaS focuses on the bottom-most layer in Cloud stack, providing direct access to virtualized or containerized hardware. In this model, servers with given specifications of CPUs, memory, and storage are made available over a network. Examples of IaaS providers are AWS EC2 (Elastic Compute Cloud), OpenStack, Eucalyptus, Rackspace's CloudFiles, etc.

Different types of Cloud Computing Providers (CSPs) [3, 4] can be classified as below:

1. **Public Cloud:** A public cloud offers its services to a full range of customers. The public nature of this model of Cloud is similar to the Internet, i.e., users and services can be anywhere on the World Wide Web. The computing environment is shared with multiple tenants, with a free or pay-per-usage model.
2. **Private Cloud:** Private Cloud services restrict their users to a select subset, usually to a specific organization within a given company. The private Cloud is similar to an intranet, i.e., their services are provided internally via an organization's internal network.
3. **Hybrid Cloud:** Hybrid Cloud providers offer their services to a narrowly defined range of private users, which if needed, can expand to reside on a public Cloud

infrastructure. Alternatively, a public service provider can remotely manage part of the infrastructure in a private organization and use Cloud for backups.

An example of a Hybrid Cloud is Microsoft Azure Stack, deployed in an enterprise but is managed externally, and if the computing requirements increase, then some tasks are migrated to an external public Cloud. This process is often called "Cloud bursting."

Clearly public Cloud presents maximal security challenges due to their relatively open access to a wider class of users. In general, when a particular model is not specified, the term Cloud computing refers to public Cloud. In all models, a client's usage is independent of the source and location of the service provider. A typical Cloud project is to develop an application which is hosted in a public or private Cloud. Thus, we will focus on planning, managing and delivering such a project from start to end.

3.2 A Business Case

In Chap. 1, we argued that every project must yield some business value. Therefore, it is imperative that a Cloud software development project too has a strong business rationale. The business case essentially brings out the basic justification of why it is important to undertake a certain project. The business case for a Cloud-based development project is prepared on the basis of scenario analysis and cost-benefit analysis. To that extent, *the business case delineates a strategic plan*. However, any major funding decision will have its own gestation period during which time the proposal passes through multiple levels of decision-making where various facets of the proposal are examined. This is done to mitigate the risks involved in arriving at a decision. The decision arrived if a justifiable business case is established. Once the decision to launch a project is taken, it is imperative that a project charter is created. The charter not only defines the scope but also delineates a roadmap of the activities along with the mechanisms to achieve the milestones. The mechanisms include the definition of the processes, choice of tools, and techniques in addition to the respective roles and responsibilities within the project team. *The project charter delineates a tactical plan.*

In this Chapter, we shall discuss both the preparation of the business case, as well as, project charter for SW development project.

3.2.1 The Information Technology Project Management (ITPM)

We will briefly revisit ITPM (earlier alluded to in Chap. 1). As a methodology, ITPM offers a framework to prepare a road map of a SW development project. Recall ITPM stipulates the following five phases [5]:

- **Phase 1**: Conceptualization and initiation of project leading to a business plan.
- **Phase 2:** Preparation of a project charter that identifies all the activities, resources available, and the actors who will be responsible for the activities keeping the priorities in mind.
- **Phase 3:** The actual implementation phase. This is the most intense part of the Cloud SW development project and will be covered in detail in the book.
- **Phase 4:** The closure of the project along with its report.
- **Phase 5:** The evaluation of the project that will help in identification of the lessons learned and identify SW assets that may have accrued from the project.

The ITPM methodology actually offers a process that ensures organized operation leading to at least one specific deliverable at the end each of the five phases. In the context of a Cloud SW development project, one would expect clarity on the following aspects of the project:

- **Project goals and scope:** These are statements that specify the objectives in terms of characteristics and capabilities of the artifact with its quality standards, in the context of the allocated time and budget.
- **The tools:** These may be either the usual support tools or some project specific tools. For example, a project to develop a Cloud-based medical application requires some imaging and graphing tools.
- **The support environment:** This is largely concerned with the Cloud HW/SW infrastructure on the one hand and the organizational support including the way the project teams are constituted and the way the teams conduct themselves.

Organizations learn and adapt as they evolve. Every project leaves its own imprint, and the organization is wiser to that extent. For each new project the goals and scope, the tools and the environment are revisited with renewed confidence within the framework of the five phases mentioned earlier. In this Chapter, we shall deal with the first two phases: creating a business plan and developing project charter.

3.2.2 Creating a Business Plan

The primary purpose of a business plan is *to offer rationale that justifies funding of the SW development project.* The senior management studies the business plan and determines if strategically it makes sense to fund the Cloud SW development project. Let us now explore the genesis of a business plan.

Usually someone in the top management, or a person senior enough in an organization, feels that there is an opportunity to offer a value proposition either as a Cloud-based product or a SW service. Also, the opportunity is such that it could possibly bring business value to the organization. With this as the background either the proposal is self-championed, or an active group ideates and seeks or offers a business plan. Most of the time, this is an internal exercise. Most organizations have a set internal process or a mechanism to generate business plans. The self-professed champion or the core group that actively wishes to push the business plan follows this process and prepares the business plan for the senior management's consideration. The plan ought to include existing competitors' offerings in terms of pricing and features to justify the need for a new project.

Cloud SW development projects may also be undertaken for the social good. These may offer better governance or citizen services. For example, Indian Railways created an e-commerce enterprise with arguably the largest customer base, offering a slew of services for public good. Other examples would be the land and green cover mapping exercises undertaken in public interest to preserve the environment. NGOs that promote social service for "have-nots" would also fall into this category. These cases are where the business value, in fact, is "*social good.*"

3.2.2.1 The Process

The business plan generation process followed by an organization is influenced by many factors like size, public or private ownership. These considerations determine the cultural framework of operation and decision-making. For example, the family owned or predominantly private businesses would have a different form of business plan preparation strategy than the government or publicly funded institutions. However, there are some common threads that capture the basic essence. We shall explore the major steps involved in creating a business plan.

Step 1: Formation of a Plan Team

While the brightest Cloud SW development project ideas come from individuals, its ultimate fruition happens through a team effort. An individual pushing a business plan would need the support of a team in building the rationale for the consideration of senior management or funding agencies like banks. The team may be assembled by the champion or even by the management. Either way, the set objective is to come up with a viable business plan. For our discussion, we assume the team approach. The following factors are often borne in mind while forming a team to prepare a business plan:

- The project team should ensure an adequate technical representation to determine the technical feasibility. Often it is required to show that the project is realizable on technological grounds.
- The team ought to have enough expertise to prepare a business case and even propose a business strategy. It should be possible to determine if the proposal aligns with the organization's vision and business goals. If it is aligned, then what is the additional value it brings. If it is not aligned, then does it open new avenues of business that may hold future.

– The team should have enough analytics ability to quantify the costs and benefits. These may sometimes be either in absolute terms or in relative terms as explained below:
  - What is the cost of Cloud SW product or service development?
  - What is the expected market size in terms of potential users?
  - What is the extent of reduction in cost for some internal operations?

Other value propositions may include an improved customer interface, better decision support, reduction in wastage, etc. The formation of a team helps in enhancing credibility of the proposal.

Step 2: Definition of Business Value

Organizations undertake Cloud SW development project if it earns them some form of business value. The business plan must, therefore, use some measurable parameters to show the quantum of and nature of the business value the organization would gain if the project was undertaken. Clearly, the team should choose parameters that are acceptable to the organization.

The numbers provided in the business plan should not only be computable but also verifiable with supporting logic based on known facts. These parameters are usually chosen to satisfy the following criteria:

– Something can be done cheaper: How much cheaper?
– Something can be done faster: How much faster?
– Something offers higher throughput: In which way?
– Something offers better quality: In which way?

Example 1 Let us consider a Fast-Moving Consumer Goods (FMCG) company that needs branch wise profitability computations on a periodic basis. This kind of project has a SW product with data-processing context. The product would operate in distributed processing mode. This requires then measurable parameters which translate into:

1. Larger number of transactions with better data management and deployment.
2. Cheaper and possibly faster transport arising from data compression.
3. Faster media support from a different protocol.
4. With greater reliability due to implementation of better recovery mechanisms.

Note that points 2 and 3 above can result in faster reconciliation of B2B transactions.

Example 2 Suppose the project is in an area of services like banking then it may translate into:

1. Cost per transaction being cheaper.
2. Reaching the end customer through a push technology preferably rather than supporting a pull mechanism.

3. Catering to a larger number of customers.
4. With clearer messaging on a smaller footprint device like mobile phones.

All the indicators are translated in terms of monetary units—be it Dollars, Rupees, or Euros or larger social good.[1] Clearly, this entails using some estimation methods [1].

The main point that must be borne in mind is the *likely impact the project may have*. To that extent some of the tangibles may not be measurable in monetary terms. The impact then would have to be spelt out in the business plan and some business value be attributed to it. Some relevant *areas of impact* to establish usefulness of the project are:

(a) Strategic.
(b) Related to customer relations management.
(c) May contribute to socially relevant issues like healthcare delivery or safety of individuals.
(d) Operational efficiency.

We give some more examples here.

Example 3 For a bank implementation of core banking solution (as a product) is both strategic and imperative for its operation:

1. Footprint: It can help the banks increase their footprint and expand their business.
2. Internal control: Better internal control and operational efficiency.
3. Quicker and reliable inter-branch reconciliation of transactions. This leads to enhanced customer satisfaction.

Example 4 For a service organization engaged in delivery of consumer electronics devices a software solution that helps them *track and deploy* the personnel may be a good solution. They can buy such a SW service from vendors under license and implement the project for a strategic gain. This can result in:

1. Serving more customers.
2. Better operational efficiency through shorter response times.

Example 5 The Government of India (GoI) had launched a project to allocate unique identification number (UID) to all Indians. This is akin to Social Security number and associated services in North America. The project stipulated maintaining a large UID database. The card is called *Aadhar*. Its primary use is for authentication. The system uses multi-factor biometrics like multiple fingerprints and iris. Authentication helps in monitoring welfare programs much more reliable. For example, it helps to monitor if the real beneficiaries do receive benefits that accrue to them. The benefits include minimum wages, minimal guaranteed employment,

[1] In the Himalayan country Bhutan bordering India and China, the ruler Jigme Singye Wangchuk stipulates that projects undertaken ought to contribute to Gross National Happiness (GNH), i.e., growth that enhances GNH.

minimal healthcare delivery, and insurance, i.e., there is a large social cause being served. Clearly, the impact is different from the usual commercial business values.

Finally, we should like to observe that the business values may be stated in absolute or relative terms like x monetary units per year or in percentage terms like 20% saving in operational cost, or with expected values like adding close to 220 customers every quarter, i.e., it should be in terms of discernible numbers.

Step 3: The Chosen Alternative and Its Justification

This is a crucial step as it *demonstrates that the alternative chosen is basically sound.* The comparison includes the base case (which maintains status quo) scenario. All viable alternatives are explored and compared. Change is desirable when the chosen alternative clearly stands out among all the alternatives—including the base case.

Example 1 To justify the core banking solution, for instance, one may argue along the following lines: Suppose we have a customer who lives in a large metro and has moved his residence. If he deposits some amount in his account in the bank branch near his new residence, it will not show in his account till the branch where his account was opened reconciles and the database at that branch is updated. This may take between 3 and 5 days (in case there is a weekend in between). With the core banking solution, the update is immediate and accessible from all the branches of the bank immediately as the deposit is recorded, i.e., within minutes. This is a comparison with what if status quo was maintained. Note that the core banking would require a central pool of servers and may be organized to operate in the bank's private Cloud.

Example 2 We shall again use our *track and deploy* solution. We describe three modes of engagement:

1. Pay per use in SaaS model: The SW service may be made available by giving an online access to modules. Every access would be chargeable then.
2. A fixed period license fee which is renewable: The SW service may be made available by giving a non-modifiable SW using a dongle or with an upper limit on number of simultaneous users. The license may be renewable.
3. Outright purchase: The service company may buy the SW with a onetime payment and hope that the cost shall be recovered with some benefits in the end.

There are other alternatives also possible but for now these three should suffice. The business plan must evaluate the year-on-year expense and projected returns on the investment to arrive at the most profitable proposition. This may be based on an analysis of the likely growth prospect in the customer base and their ability to support the service.

Step 4: Assessment of Risks

Whenever a change is contemplated, an assessment of risks is warranted. The risk may be evaluated from several angles such as technological feasibility, organizational acceptability, and cost consideration. There may be other factors such as the safety, legality, geo-political consideration, and social impact. In fact, all the above-mentioned considerations apply in the context of outsourcing.

Obviously, the very first step is to identify risk(s). Once a risk is identified, one becomes aware of what may go wrong. Should the worst fears come true, then what may be its impact? This needs analysis to determine the consequences. Finally, of course, we also need to ask which way to mitigate? Briefly, the risk assessment involves three steps: *identification of risk, its impact and consequences, and finally a mitigation plan.*

In Cloud SW development projects, often the main concern is about the features in the final artifact. So, one may proceed on the basis: *what must not go wrong and what may*. Core features must always be provided for. What, however, may go wrong is when some feature is either partially implemented or does not account for some operational scenarios. There may be a platform or interoperability-related issue, such as browser or API compatibility.

We discuss an example here that brings out two forms of risk. The first one relates to technological feasibility while the second deals with organizational acceptability.

Example Hypothetically, consider a project for Cloud-based gaming software development. The platform should be comparable to Sony's play station. Sony and IBM had earlier collaborated to develop games SW around IBM's Cell Broadband Engine (Cell BE). Cell BE had a full SW development kit and a library that supported huge number of computational geometry algorithms for realistic rendering and real-time response to human interaction. So, one possibility is to use Cell BE, whereas the other possibility is to use new multi-core platforms. However, the choice may be from Intel, AMD, or ARM. More so for a game engine SW, we need a software development kit along with some debugging tools. As of this writing, there are some HW debugging tools like CoreSight[2] but hardly any for Software development activity. This is especially true when gamers need to interact with each other using the Internet, then concerns of security and performance also creep in.

This business case also touches on some aspect of organizational acceptability. The management may need to decide to buy a Cell BE System [6] on Chip (SoC) and use it. However, buying may become a contentious issue if some employees believe that they can develop an in-house solution around some multi-core platform. The management may not wish to risk delays that could accrue from an in-house development effort. It may even be wary of the reliability of the internally

[2] https://www.sciencedirect.com/topics/engineering/coresight-debug-architecture, uses two pins: a serial-in and serial-out to provide hardware debug and trace support.

developed module. Note, however, if maintainability is a major consideration, then in-house development may after all be a better alternative. The team needs to do a make vs. buy analysis, keeping in mind both the development and maintenance considerations over the life of a product. Another consideration is a change in development staff, e.g., if some key engineers leave the project, then their work needs to be well documented and picked by other team members. Similarly, using SOUP[3] (Software of Unknown Providence), not excluding open-source software brings external risk factors such as dependence on third parties to fix bugs and secure the code. In this case, the risk can be minimized by picking components that are well used by others and maintained by a community of developers. The process to identify such components within a company is called whitelisting.

It is said that hindsight is 20/20, which refers to a perfect vision (e.g., an object 20 feet away appears to be 20 feet away) looking backwards in time. Thus, it is interesting to note that Sony Playstation started to use the Cell BE project in mid-2000. Their first such product was announced in August 2009 with the launch of Sony's Playstation 2, but by November, 2009, IBM had discontinued the development of a cell processor [6]. Perhaps this was due to a make vs. buy consideration as IBM's specialized designs had difficulty keeping up with the performance of GPGPU (General Purpose Graphics Processor Unit) architectures from NVIDIA and AMD. The then popular Sony Playstation 5 design used a SoC made by AMD, while the gaming software has shifted to Cloud [7]. The Cloud service lets players play the games without downloading the software on a Playstation (PS), including the ability to play certain games on a PC. PS is now a subscription-based service that grants access to a large library of games. Players are given a slew of options ranging from monthly to annual subscription. The processing tasks are performed by remote servers that render videos on the local screen, and quality of images depend on the bandwidth of Internet connections rather than the local processing power. This shift away from special purpose cell to commodity hardware and from local software to Cloud-based games indicates the lure of Internet as distribution channels are simplified. This trend will continue in other domains too, as business models shift to using Cloud for providing software and services.

Risk assessment commonly use *scoring methods of quantitative nature* and this will be covered later in this chapter.

Step 5: The Cost-Benefit Analysis

No enterprise would invest in a software project unless it is perceived to bring some form of tangible or intangible benefits. Some organizations attempt this exercise by computing *cost of ownership* and *benefits from ownership* as the measures. The cost-benefit analysis—with benefits outweighing the costs—would tilt decision of

[3] https://www.softwarecpr.com/2020/10/understanding-ots-and-soup/

the management in favor of executing the project. The business plan offers an exposition of the cost and benefits in quantitative terms.

As for the cost, it comprises the following major components:

1. The fixed cost non-recurring infrastructure: This is the most obvious one-time cost that has to be borne any way. It covers IT infrastructure costs like hardware and cost of communication infrastructure besides the other tangibles like office building.
2. The non-recurring human resource cost component: This covers those who may be hired temporarily as consultants. Also, it would include any one-time cost that covers specific training requirements.
3. The direct recurring cost: This cost covers the disbursement on a regular basis. It may span the entire lifetime of the project. It may have followed two components.
   (a) The equipment or software license and maintenance fees, utilities like power, etc.
   (b) The project employee salaries.
4. Indirect costs: These are also one-time costs but arise from factors like costs due to some kind of temporary dislocation, loss of productivity during the cut-over period, and possibly other unavoidable procedural considerations.

With the costs identified, one can compute the *total cost of ownership* (TCO) by summing various cost components as defined above.

For a full cost-benefit analysis, we also need to indicate the likely benefits. The benefits have to be associated with business values which can be quantified. While we shall describe the quantitative methods later, it would be worthwhile to list the forms of benefits from the SW development projects. Typically, the following forms of benefits are easily identifiable:

1. Raising the productivity level of employees.
2. Fewer operational errors with better and quicker updates.
3. Better control and communication support for decision-making.
4. Raising efficiency within the enterprise.
5. Better response to customer needs.
6. Quicker response to programs for social good.

It should be remarked that the cost-benefit analysis is ultimately translated to some quantifiable number in currency unit which may be in use. The total computed from of benefits can be used to compute total benefits of ownership (TBO). When the TBO exceeds TCO, then an internal Cloud SW development project is viable.

A recent example of this is Uber. It is a Cloud-based application that connects drivers to riders, using smartphone as a tool [8]. Starting in 2009, this company operates in more than 900 cities worldwide with a valuation of more than $70 billion. It does so without owning taxis or employing drivers. Its business model is commission based, as a rider indicates the need for transportation and the Cloud application connects this need with several available drivers in the area. Based on a mutual selection and agreed upon price, the ride takes place. It can be shared by the rider for tracking by other third parties to ensure compliance and safety. Drivers get paid on the basis of rides they deliver, while Uber takes a percentage share of each ride. Main attraction for the riders is flexibility with fares cheaper than taxis. In larger cities, many inhabitants no longer own their own cars as Uber rides may cost less than TCO of a car. At a higher level of abstraction, Uber offers an aggregation platform that ensures matching provider provisioning of a service (i.e., drivers with cars), with the consumers (i.e., riders). Similar business models are used by Netflix, Airbnb, and even Amazon with third-party sellers to connect producers of a service with consumers of that service. Such platforms are enabled by web services in Cloud. Previously, specialized applications needed their own compute servers and storage locations, such as Ebay. However, due to a large variability in consumer demand at different hours of the day, or different days of the week or different periods in a calendar year, any dedicated compute facility would be either oversubscribed or underutilized most of the time. In Cloud, the demand variability is addressed by resource sharing and intelligent scheduling of tasks.

Step 6: Comparison of Alternatives and Support for the Final Recommendation

A business plan supports a strategic decision: Whether to go ahead with a certain SW development project. Therefore, it must provide a sound basis, i.e., without leaving any scope for doubt. It is quite possible that there are many viable options where benefits exceed costs as defined in step 5. However, one would want to *know if the alternative scenarios have been explored in depth and compared to support recommended strategy.*

The comparison of alternatives should minimally indicate the financial basis. It is important to look at cash flow management. This may be in terms of capital investments and borrowings and modes of repayment. In simple terms, the costs define the outflows, and the benefits define the inflows. The alternative chosen should offer a viable and manageable cash flow.

One may also explore the payback period. Invariably an investment in any project should be accompanied by an analysis that projects how the investment made will be paid back over time. For example, the analysis may indicate year-on-year return on the investment. There may be ways in which this may be computed.

The other considerations may have longer term impact analysis in being able to quickly upgrade technology or adapt to an emerging technology for maintaining continuity of the particular line of business. It is also possible to see if the business helps in spawning off other tangential activity with some potential benefits. A case in point is the way the mobile ring tones (as data) have influenced the music business.

With a uniform basis of cost-benefit analysis for each alternative, a final recommendation may be attempted. However, it is quite possible that there may not be a common or a uniform basis for comparison in absolute numeric terms. However, it may still be possible to make a *relative comparison*. The quantitative methods used for relative comparison is to use a *scoring method* over some scale like high, medium, and low or spread over categories “A” to “E” and then later assign some weight to each category. The scoring method is often resorted to when alternatives do not have commonality of parameters. That leaves no choice but to categorize the alternatives and map these to some perceived parameters of goodness by associating a weight with these. In fact, this is how the student’s relative standing is obtained by using their performance over different subjects. This final step also must list out the option chosen with accompanying recommendations.

We have listed out six key steps in the preparation of business plan. However, depending upon the nature of the SW development project, the report structure may differ. It may have sections highlighting aspects that are specific to the project. For instance, it may have a section rationalizing the choice of technology or a technology partner selection. However, it is still possible to offer a template to draw up a business plan. In fact, a simple search on web will yield business plans for any chosen domain. The main advantage of using a template is that it helps in organizing the report. Besides it is likely to align with the expectations of the management—at least in the structure. The senior management can efficiently explore a structured report.

In summary, a business plan document has a formal structure which details vision, operational road map, and the nature of financial support. For it to be useful, it must cull out the business value proposition. Appendix 1 gives a sample structure for a business plan for a SW development project.

Let us now explore the software project charter description which includes a project plan.

3.3 Cloud Software Project Charter and Planning

The Cloud SW project charter and planning documents may be considered as the most complete source of information on every aspect of the project. These documents are the ultimate point of reference during a project’s lifetime. These documents lay out road map for evolution of the project. A SW project charter usually captures the scope, budget, deliverables, infrastructure, human resource, administration, the risks, quality guarantees, acceptance criterion, and the project closure protocol.

The project charter may either include or be supported by a project plan document which covers the schedule with identified deliverables. The plan document identifies project tasks with time estimates and sequencing order. The project’s organizational structure and human resource management find a prominent place in the charter. The idea is to show how SW project charter *provides the tactical plan that guides the execution* of the project. Our goal is to discuss desirable characteristics for the charter.

In Appendix 2, we provide a template to help prepare a Cloud SW development project charter.

3.3.1 Vision Statement and Overview

The project charter opens with a vision statement and a brief overview of the project. The vision statement is either reproduced from business plan or may offer some elaboration. The primary purpose of a vision statement is to keep the focus through the project's lifetime. The vision statement becomes the guiding beacon reminding people to make choices when faced with the challenges.

The vision statement encourages *project leaders with cognitive understanding to ideate* to realize the project's goals. As Scott Berkun [9] puts the perspective of vision statement should be "*inspirational*" and "*intentional*" at three levels given below:

1. What I am doing (at the individual level)?
2. What some of us are doing (at the project segment level or small team level)?
3. What every one of us is doing (at the project team level—as the overall goal)?

Following the vision statement, usually an overview of the project is presented. The overview of the project identifies all the stakeholders with their respective roles. It also describes the objectives and resources available. It also lists all the deliverables. The deliverables description leads to both functional and non-functional requirements—like the expected level of performance and quality attributes including reliability. In addition to a brief on the scope and budget, the overview provides the schedule and may also touch the project management structure briefly.

The key parts of the charter describe scope, budget, deliverables, and the plan along with the logistical support by way of infrastructure and human resource. It identifies deliverables with quality guarantees, acceptance criterion, and finally the closure protocol. We next discuss each of these in some detail.

3.3.2 The Scope and Deliverables

Often the scope is expected to define both *what it will deliver* and *what it may not*. From a technical stand point, we would say *it defines the extent of capabilities the project shall offer to the end user*. The scope provides the foundation for the preparation of budget as well as schedules. A mid-way change in the scope is often the reason why projects fail.

As examples of the kind of statements, we may revisit the three examples cited earlier: core banking, consumer electronics service provisioning, and the UID.

- For the core banking project, the scope may include the following:
  - The bank shall provide online third-party fund transfer facility to savings account holders.

   - The banks shall permit only account status queries on the mobile for current account holders.

- For the consumer service provider, the scope may include the following:
   - Following an online registration of a complaint—a service engineer shall attend to it within the next 24 h.
   - If the gadget requires a spare part, the same shall be provisioned within 2 working days from the OEM sources unless the spare part production itself is no longer supported.

- For the UID project the scope may include the following:
   - The UID shall have biometric identifications, namely the finger prints and iris scan.
   - The UID will automatically register the potential beneficiary with welfare program to which the person is entitled.

It is important to note from the above examples that *the scope unambiguously defines what the project must deliver.*

Deliverables The charter spells out the Cloud SW deliverables in its entirety. The deliverables are often expanded to define the functional and non-functional requirements—like the expected level of performance and quality attributes including reliability. To get a perspective on how deliverables may be stated, we give some indicative deliverables for the three examples we have been taking forward.

- For the core banking example, some of the deliverables could be:
   - Account opening with a unique account number. The account number should identify the branch from where the account has been initiated.
   - Upon closing an account, the entire accounting history shall be archived. This account shall be frozen, and this account number shall not be reused.
   - There will be account reconciliation between the bank branch database and the central data repository on a daily basis.

- For the consumer electronics service center example, some of the deliverable could be:
   - A spare parts database shall be provided for inventory management. For each spare part, the database will have a predetermined reorder level. Whenever a certain spare part inventory falls below its reorder level, an alert shall be raised to order the spare part or component.
   - A profiler shall be provided to profile the service engineers. The profiler shall maintain the customer satisfaction rating of service engineers. A monthly report shall reflect the performance of individuals. The performance shall be measured in terms of number of customers serviced and their rating of the service quality.

– For the UID project, for example,[4] deliverables could be:

- The card shall be RFID compliant (this actually means it shall have an embedded IC chip).
- For authentication, the card shall not only carry photo identification but also biometric information for the finger prints and iris scan.
- The first four digits in the UID shall identify the location where the card originated.

The deliverables stated here are only representative. Each software project is bound to have some its own characteristics reflected in the deliverables.

Graham McLeod and Derek Smith [10] have proposed a definition structure for deliverables. A table structure is shown in Table 3.1 below based on their definitions. The reader may specifically note the column headings in the table. The heading completely characterizes nature of the deliverable. As an example, we show the characterization of inventory database for the example on consumer electronics service center.

In Table 3.1, we have mentioned a database standard SQL 2019 or later, the requirements emanating from these standards may be more explicit. For example, one may require that the database minimally meets the requirements for page locks in addition to the Atomicity, Consistency, Isolation and Durability (ACID) properties.[5] There might also be usability considerations in terms of user interface and access provisioning from mobile phones for instance.

There are times when a project scope may need to be revised or deliverables may be modified. This may be due to several reasons. It may be some form of exuberant optimism, incorrect set of assumptions made, or uncertainties that could not have been foreseen. Therefore, a provision should be made for scope and deliverables to be adjusted during the lifetime of a project. In the event of having to resort to changes in the scope and deliverables, a proper log needs to be maintained. The project management tool suites provide for incorporating such changes.

Table 3.1 Deliverable statement for consumer electronics spare parts inventory management

Deliverable	Structure	Standard	Ownership	Approval by	Resources
Inventory database	RDBMS	SQL—2003	Database lead for local DB	Project manager	Database server
Customers unstructured feedback	NoSQL	No standards exist	DevOps in the cloud DB	Project manager	MongoDB
………	……..	………	……………		………..

[4]This is only an example and bears no resemblance with the actual deliverables used by UID authority in India.

[5]It is well known that it not possible to support ACID properties in its entirety. However, most DBs work around it.

3.3.3 The Organizational Structure

Cloud SW projects are time-bound in nature. This entails that there is clarity on the roles and responsibilities of individuals. It should be clearly known who is entitled to authorize what. This helps in quicker decision-making across all levels within a project. SW project charter, therefore, must clearly indicate the types of roles and responsibilities with an appropriate reporting structure. Broadly this includes the responsibilities of project administration and budget control, SW development team formation and recruitment, project monitoring, system configuration and change control, quality assurance protocol, client interface management, and project documentation management. These responsibilities need complete clarity within a project's operational frame.

Generally, it is true that a single point of authority is vested in the project manager and that person will create his/her own framework. However, project management is team work. In practice, considerable operational freedom would be given to key individuals in their spheres of activity—if only to ensure success of the project. For instance, the quality assurance manager and SW designers would have considerable freedom in choosing their programming team members.

In Table 3.2, we show roles, associated responsibilities, and a typical reporting structure. Table 3.2 basically suggests a hierarchical structure. However, in practice, there is considerable peer level communication and cooperation. For instance, testers and developers or programmer teams need to communicate. Similarly, the SW architect or designer manager would need to communicate with quality assurance manager. Also, admin would need to communicate with accounts too. In Fig. 3.2, the dotted ellipses capture some of these the peer level communication. In practice, the charter would provide for a separate communications management protocol as discussed later in this section.

Notwithstanding our description of the organizational structure in Fig. 3.2 and Table 3.2, there may be other considerations that may determine SW development project management organization in specific contexts. Some of these are given below:

- The basic philosophy traditionally used within the organization.
- The size of the organization.
- The nature of the project.
- Some very specific requirement from the client.

To go over some of these, let us consider a large organization that is involved in multiple product and services. Invariably such organizations have a *functional division*. For example, organizations like Wipro and Infosys have distinct verticals like testing, infrastructure to support domains such as banking. When a client approaches them to set up something—say a core banking project—then each of these functional divisions provide a few personnel, who are regular employees, for this project. For that duration they report to the project manager even while the overall administrative control of personnel remains with the verticals they belong to. In fact, the functional division suits larger organizations from operational point of

Table 3.2 Organizational structure for SW development project management

Role	Responsibility	Reports to
Project manager (P.M.) or engineering manager	Overarching responsibility for the project, ensure that resources are made available as required to deliver the project on time and within budget	Board member/ management
Consultant	May be internal or external, for example, a Linux kernel expert In the organization may act as a consultant. In banking context, a professional banker may act as an external consultant	May report to Mgmt. board/P.M.[a]
Analyst	Do system analysis of what is required by the customer, provide estimates in terms of effort required, explore options for the choice of environments and platforms and its impact on schedule	P.M.
Architect and designers	Define the overall SW architecture with modules, interfaces and how they may be developed. May also define detailed file and data structures	P.M.
Quality assurance manager	Define quality attributes and create a test schedule for functional and non-functional tests (performance, load test, defect mgmt.)	P.M./SW architect
Program leads and test leads	Based on design details work out a deployment schedule of programmer (tester) activities (to meet the schedule)	SW architect/ QA manager
Programmers and testers	Based on the details worked out by program lead (test leads), write the programs (test programs)	Program lead/ test leads
Documentation/ report generators	Keep a documentation and prepare reports including maintaining minutes for the project monitoring	P.M./SW architect
Client interface	Collect user requirements, communicate progress, encourage their participation in monitoring, testing, workout project handover	P.M.
System admin	Manage the system support, update, installation and configuration management, support change management	SW designer/ architect
Library	This is to preserve programs and utilities for later use. It may also support archiving	SW designer/ architect
General admin and legal support	Provide administrative support (HR services like recruitment, legal support, travel, leave records, secretarial services, etc.)	P.M.
Accounts, stores and budget	Provide accounting support (salaries, procurement, budgetary control, general expenses, etc.)	P.M./Chief of finance
DevOps engineer	Guides software developers to make code releases and helps with deployments in the cloud	P.M.
Scrum master	A facilitator for an agile development team	P.M.

[a] P.M. stands for project manager

view. Each organization may tweak the method to suit some of their traditional way of working. The function-based division also leads us to a matrix form of organization as shown in Fig. 3.3.

In the context of Fig. 3.3, it should be remarked that architects, consultants, quality assurance managers, and system configuration managers may be involved with

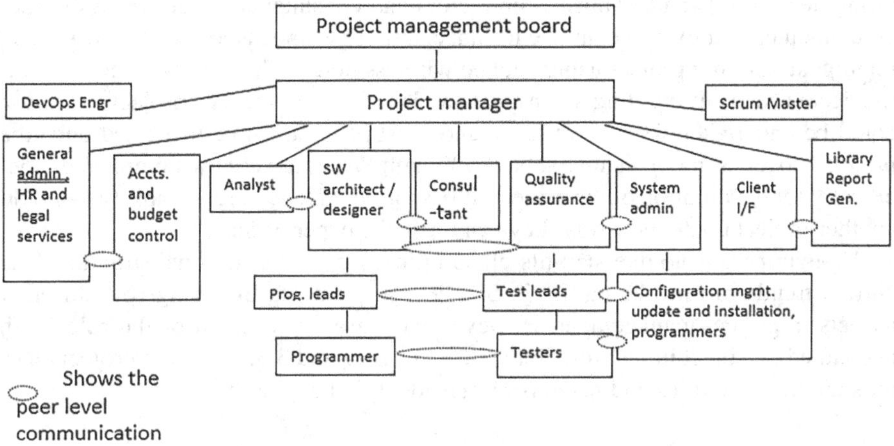

Fig. 3.2 Organizational structure for SW development project management

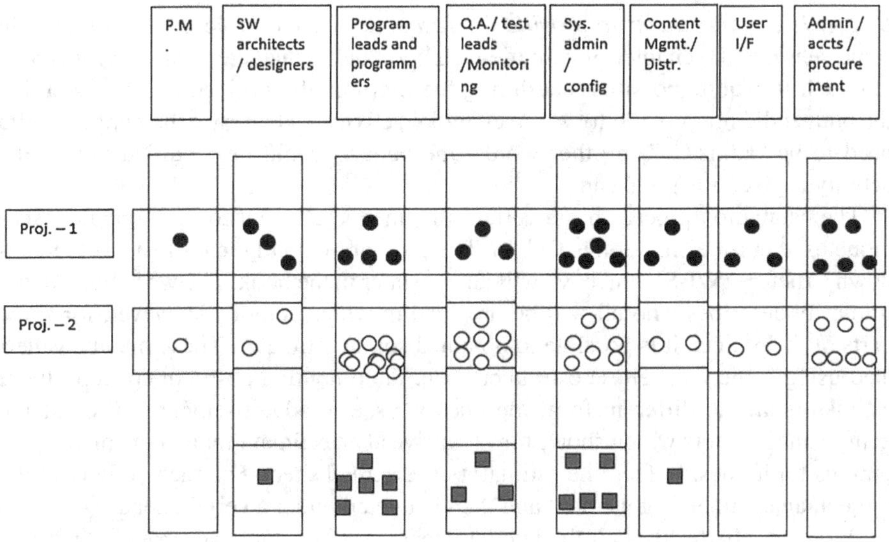

Fig. 3.3 A matrix for SW development project management

multiple projects. For example, a multi-national in Bangalore maintains individual account for each project—but also maintains a few competency pools. The pools offer need-based support for individual projects. For example, they have a strong open-source Linux group primarily to support any Linux-based project. Also, one Linux lead may get involved with more than one project.

Sometimes smaller organizations (usually small and medium businesses) assemble a crack team to execute the project. Usually, the embedded systems and media-based project(s) get done that way. The nature of some project may either require hiring new experts and/or hiring contractors and consultants to deliver on the specific modules. For example, in one instance, a small team was formed to work on an intelligent building project using Zigbee wireless nodes which between them communicated relevant building's sensor data. This kind of specific project operation could be undertaken for a short term, and most of the team is disbanded once the project is over. Sometimes up to about 50% employees are contract employees specifically for the duration of the project. It has the advantage of repeatability—should another project come their way they can reuse the experience.

Though there is no one-size-fits-all solution to the organizational structure. The current trend is to invoke agility in SW development. Agility brings out different aspects in project management. However, we can identify most of the roles with associated responsibilities from Table 3.2, Figs. 3.2 and 3.3. A typical project charter shall have the roles and responsibilities identified explicitly.

3.3.4 Project Budget and Schedule

Budget is a commitment to provide for resources for the project [10]. Budget, in some sense, is the enabler of activities. Only on being budgeted, an activity can be initiated. It is quite possible that during the lifetime of a project, it is realized that the budget did not provide for some essential activity. In that case the budget would need to be "adjusted." In other words, the budget would be modified before the activity is given the go-ahead.

The estimation process has a certain judgmental element to it. A popular technique used in for estimation is Delphi. The estimation finally leads to a work breakdown structure (WBS) which we will take up later in the book. The work breakdown primarily describes who all will be involved in which activity. However, for some parts of the project, it is possible to obtain absolute estimates. These may be generated using some tool. These estimators help in computing costs. In concept, these estimators are no different from the ones we see in advertisements for emulsion paint manufacturers which shows how to arrive at an estimate for a range of painting options for homes. In fact, the estimators use spread sheet. For each activity in the project, an estimate is generated and then budgeted under a certain head.

A project charter lists all the budget heads based on some accepted categorization. For example, there would be a head that accounts for fixed cost, recurring expenditure which include some indirect costs, contingencies and reserve funds, learning cost, etc. Below we give some representative indicators on what the budget heads may provide for:

- Fixed cost budget may cover:
  - Equipment.
  - Infrastructure.
  - Software purchase or one-time license fee.
  - Contractor services etc.

– Recurring cost budget may cover some direct and indirect costs, such as.
  - Wages.
  - Regular supplies.
  - Annual license fees, insurance (indirect cost), etc.
  - Administration (indirect cost).
  - Travel (indirect cost).
– Contingencies and reserve funds budget may cover:
  - Emergent expenses on travel (contingent expenses).
  - Replacement of parts (contingent expenses).
  - Reserve funds (to cover risks).
– Learning costs cover the following:
  - Rapid prototyping costs—the ones that shall be thrown away.
  - The cost of training personnel.

Budget monitoring is an important activity, and project managers maintain a check list to monitor the expenditure under every head of the budget. The charter may schedule a budget monitoring regime as a part of project review. The section on budget in the charter would also identify the chain of authority for any approvals and purchases, revision, and even write-offs.

The budget also *sets expectations*. The problem is that in most cases, the budget is set at a lower level, while the expectations are established at a higher level. Invariably there would be cost overruns for a variety of reasons. Suppose, for example, if one estimated that the project shall cost between say USD 10 and 12 million, then the senior management would be averse to any upward revision beyond USD 10 million, while the expectations set may require USD 12 million or even more.

Project schedule is often tied to the budget. As we mentioned earlier, budget is the enabler of activities. Almost always the budget allocation is spread over the project's lifetime, i.e., not all budgetary resources are available from the beginning. The budgetary releases are schedule based on the expected (or evaluated) rate and extent of progress. Also, note that the activities may have some precedence order, i.e., some activities can only be undertaken when some others have been either been completed or in various stages of completion. We can explain precedence possibilities by imagining that a certain project has activities P and Q with the general precedence defined by P precedes Q. Andrew Stallman and Jennifer Greene [11] mention the four kinds of possibilities that can occur:

1. **Finish to start dependency:** ***P*** must finish before ***Q*** can start.
2. **Start to start dependency:** ***P*** must have started before ***Q*** can start.
3. **Start to finish dependency:** Till ***P*** starts ***Q*** cannot finish.
4. **Finish to finish dependency:** Till ***P*** finishes ***Q*** cannot finish.

The key point to note is activities ***P*** and ***Q*** could be overlapping in time in the cases 2, 3, and 4 above.

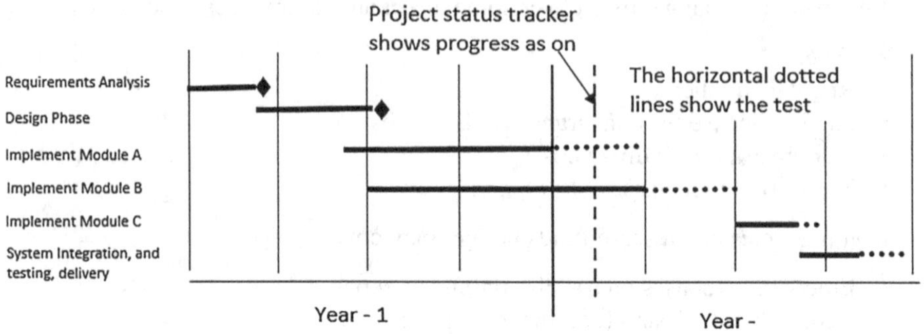

Fig. 3.4 A scheduling example for SW project

There are tools that help preparing a project schedule with budgetary control. The tools trigger resource release upon the completion of some activities. These tools may use Gantt charts, Project Evaluation and Reporting Technique (PERT) or a Fish-bone diagram. We will discuss this topic in an Appendix 3. For now, we will assume a simple scenario. In Fig. 3.4, we show a simple Gantt chart. It shows a 2-year project timeline over 8 quarters covering activities to develop three modules: A, B and C with C dependent on completion of modules A and B.

In the figure, note that module C is scheduled after completion of modules A and B. An activity's time-span is show as a line. The diamonds of a line denote termination of that activity. The design phase starts only after the requirements phase is finished. On the day, the tracker shows that the requirements analysis and design phases have been completed. It also shows that Module A and B implementation is in progress. Module A is in the testing phase. Module C has not started yet.

The chart such as this one can also be used to reflect upon how one is progressing with respect to the schedule. In fact, the project managers are expected to use check lists to ensure that they effectively monitor the project's progress. It is also important to reckon that budgets and schedules may have to be revised in the case of unexpected events. This necessitates a project management tool that can track the progress made. In cases where some changes are needed, then these are incorporated in the project on a real-time basis. Some key features of a project management tool are:

1. **Project Estimation:** This helps to establish a baseline and build budget forecasts.
2. **Budgeting:** This is needed to track expenses, track actual costs to invoice clients and pay employees, and compare planned and actual costs, managing costs to stay within the planned budget.
3. **Resource Allocation:** This is needed to ensure that work is assigned based on the skill sets and availability of resources, to avoid over or under work.
4. **Collaboration:** This is to keep all project stakeholders on the same page. With a project split over many geographies, web-based chat, meetings, and forums are needed to achieve communication and consequently the cooperation.
5. **Quality Management:** This is to ensure that all issues are logged, tracked, and resolved in a timely manner.

3.3.5 Project Plan Document

A detailed project charter could include all the major aspects related to the project plan. At the very least it would list all the major *milestones* for the project. Sometimes the project plan may be brought out as a separate standalone document. In that case, it would have to be a very detailed plan. The plan document covers the scope of all the activities in the project.

The plan covers scheduling of tasks and deliverables, human resource planning, and deployment. In short, it would include an abridged work breakdown structure (WBS). The charter should have a timeline for project monitoring, review, and reporting method as part of the communications management strategy. Monitoring and review would include quality assurance plans as well. Yet another very important part of the SW development project plan is risk mitigation. Risk mitigation happens with a reality check and is followed by revision or fallback options. This happens when some of the assumptions need to be revisited in the context of uncertainties.

To illustrate the nature of information that may go into a SW project plan—let us carry forward our example of the company that engages in consumer electronics repair service. Suppose the project now requires that the service is upgraded to a mobile SMS application.[6] The enhanced scope then shall offer the following additional features:

- Customers may use SMS.
- Manager may use SMS for deployment and collaboration.
- Service engineers may collaborate using SMS to reroute themselves to service the closest customer.
- All SMSs are structured to be database compliant.
- Database alert should be SMS compliant.

Briefly, let us elaborate some salient features of the Cloud SW project plan shown in Fig. 3.5. We also indicate specific *milestones* with the motive to highlight how the project plan or charter document should be organized to identify the tasks or sub-goals that lead to the final realization of Cloud SW project.

The plan will detail how the resources like budget, project personnel and end-user training, and mobile service provider selection will happen. The schedule would show updates for completing different phases of the project in 6 months' time.

- The duration is 6 months with close monitoring every 2 months.
- The initial budget release is high and then it tapers down. Towards the end, it peaks to finally resolve all the pending financial matters. We have shown allocations in percentage terms. However, in actual plan documents, the numbers would reflect amounts being released. Also, note that larger amounts are released

[6] One of the first applications was to manage gate allocation at the airport to manage aircraft movement at the air ports. An Introduction to *Operating Systems*: Concepts and Practice; Author, Pramod Chandra P. *Bhatt*; Publisher, Prentice Hall of India, 2003; ISBN, 8120323068

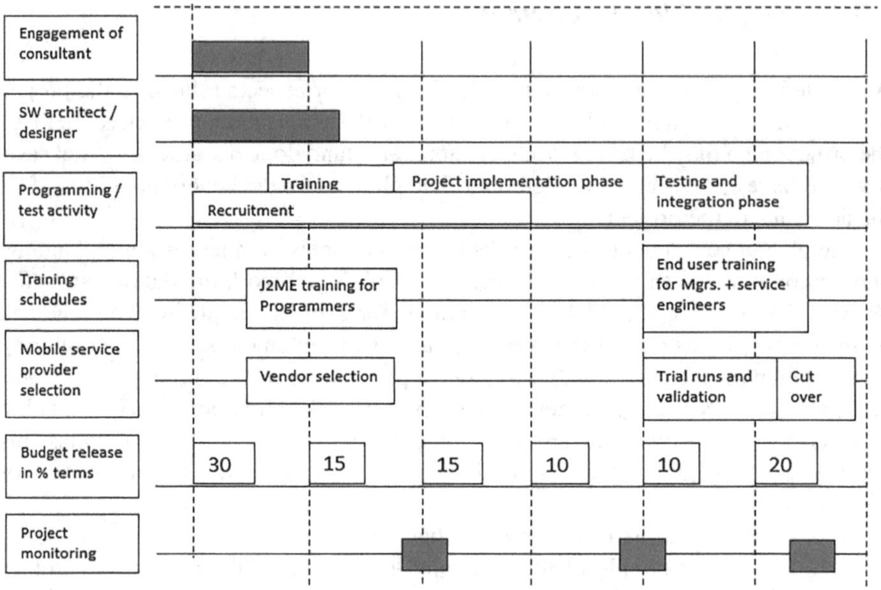

Fig. 3.5 The schedule for the project on consumer electronics mobile-enabled service

at the beginning and towards the end because quite often some investments towards infrastructure or equipment may need to be made at the beginning. The recurring cost is relatively uniform. Towards the end, again a proper financial closure may require additional allocation.

- The project involves vendor selection. The plan must have time-table for it. The completion of the process of selection of vendors is a milestone for a project. This applies also when a contractor is asked to develop a specific SW component which will later get integrated. Similar selection procedure is needed to select a third-party SW component. Also, note that the trial runs may involve cooperation and participation by the service provider.
- The Cloud SW architect and designers analyze the requirements.
  - Identifying all the modules that would be needed.
  - The manner in which the system would be integrated from the modules.
  - The architect may also indicate the likely effort required for completing the project.
- The recruitment phase is spread out. This is because it is planned to recruit protocol experts and testers just before their services are required. We have made reference to this aspect later.
- The plan should specify the cut-over period. The completion of cut-over is also a milestone.
- The recruitment is usually at multiple levels. So, each number should indicate different roles at each of the levels. Completion of recruitment is a mile stone so

also completion of their training period. In fact, the implementation phase may have several mile stones for various modules as well as when an integrated package with multiple modules is ready. If the organization follows a development testing methodology, then the phases would include testing of individual features as these become available. Addition of every fully developed and tested feature shall also be identified as a milestone—at least as a minor milestone.
- The project stipulates training of the end users.
- The consultant's help is required in the initial part of the project to ensure that requirements are precisely defined, i.e., there is no ambiguity in the requirements. Finalization of requirements is a major milestone.
- Though not shown but it is understood here that there is a project manager for this project. He is assumed to have been identified as the project gets launched.
- The project manager may form one or more management teams for both close monitoring and reporting.
- The project manager himself would report to a project steering committee.

We have sketched the basic points only briefly. However, in practice each point would have a fairly detailed elaboration. Presently, let us look at the work breakdown structure. Later, we shall explore risk mitigation strategies and quality assurance plan.

3.3.5.1 Work Breakdown Structure

In the previous subsection, we saw a plan for a specific project. However, for a broad-based explanation, we need to recognize that every project is divided into *phases* and each phase has deliverables. When a set of *activities* in a phase lead to accomplishing a deliverable, then that deliverable is considered to be a *milestone*. The activities define identifiable work units which are logically wholesome. The logically wholesome work units may have a dependency—in that case these would have to be done sequentially. The work breakdown structure primarily identifies activity sequences that lead to the deliverables.

The work breakdown structure should certainly take into account the internal resources and the dependence on the external inputs.

Work breakdown structure for a project evolves through the following set of steps:

1. The first step is to identify all the logically wholesome units of work to define activities.
2. Define an activity sequence that leads to a deliverable.
3. Estimate the resources required for completing a deliverable.
4. Get an estimate of time and effort that is required for each deliverable leading to a milestone.
5. Map the personnel and available resources to create a deployment plan. The deployment plan may be as detailed as may be possible. For instance, it may have fine granularity in terms of man-days.

As one can see from the above, the work breakdown structure is a meticulous way of stating: *who does what, when, in which order, and with how much effort*. Note that in addition to the functional compositionality testing, quality assurance, configuration management, and change control activities also get included in the effort estimation. The key step (3) above is the estimation of the effort. There are many estimation techniques that have evolved over time. The chief among these are as follows:

1. **Expert opinion:** An estimate based on someone who has experience with project estimation.
2. **Delphi:** A method that goes back and forth between estimators. Each estimator works with his parameters and understanding initially but the estimators get to revise estimates by observing and accounting for the concerns that may have been overlooked or not understood initially. The method finally converges.
3. **Time boxing:** This method is driven by considering that each activity shall have a bounded extent in terms of time allocation. The objective being that by bounding it is more often accomplished. This is because people work with time boxed deadlines to accomplish goals.
4. **SW metric-based estimates:** These methods would include code size, function point analysis, or COCOMO methods.
5. Estimates based on top-down or bottom-up approach.

The Cloud SW development project plan should indicate the basis of its work breakdown structure. Also, the work breakdown structure should clearly identify the internal resources and the dependence on the external inputs and that includes third-party involvement or subcontracted work.

An increasingly popular way to build software products and deliver them in the Cloud is a Continuous Integration (CI) and Continuous Delivery (CD) pipeline [12]. Collectively known as CI/CD, it helps to automate software delivery process by building code, running tests, and deploying a newer version of the application. Specifically, CI is a software development practice in which developers merge code changes in a central repository. Each change triggers an automated build-and-test sequence for the project, thereby developers get immediate feedback on the correctness of their changes. CD is the automated process for infrastructure provisioning and deployment. If a certain stage fails, then appropriate notifications are triggered by email or Slack[7] to let the development team know about the cause. This enables developers to stay focused on writing code, while QA and product stakeholders can access the latest, or any earlier version of the system.

[7] https://slack.com/ is popular teams' communication tool to connect across departments and time zones.

3.3.5.2 Risk Management

A SW development project often is based on assumptions and some expected levels of subsumed certainties. However, there can be unforeseen loss due to invalidation of assumptions or uncertainties that creep in. Risks have the connotation of possibility of such losses. Therefore, some assessment of risks helps in managing the project better. Surprising though it may seem, risk management also *enables aggressive risk taking*. Project managers, who succeed, may have taken aggressive risks. If they pull through, they are regarded as very successful.

For now, we pay attention to the scenarios when the assumptions may not hold or uncertainties creep in. Either way the project should have a risk management plan embedded in it. Here are some forms of risks that may show up:

1. Most projects employ some SW metric to estimate the effort required to implement the project. Effort estimate is usually related to the expected size of the code (lines of code (LOC)) and the complexity of functions that need to be coded. There are many SW metrics [13] defined to obtain a measure of the effort. Ultimately the effort translates to *person months* required. The effort estimate helps in determining the number of people required in different stages of the project. It also helps in determining the time required for completing the project. If the effort estimates go hey-wire, then minimally there is a risk of delayed delivery. It may even result in having to abandon the project itself.
2. The schedule is prepared with some assumed levels of productivity of the personnel engaged in the project. In practice, every project has an initial intellectual phase when the people feel challenged and are prepared to put in a lot of time and effort. The productivity is high during this period. In later stages of the project, productivity of the same people comes down [14]. In addition, the productivity assumptions across the workforce may have a wide spread (very high variance statistically speaking). This can result in much poorer performance than expected. This too can result in terms of delay, cost escalation, and even poorer reliability of the artifact.
3. One of the main concerns in Cloud SW industry has been: the people turn over. Sometimes people show poor professional integrity and ethics by leaving their project at some arbitrary point of progress. Departures should always happen after some logical milestone has been achieved and documentation prepared to enable proper handing over of the project. Often when a key developer leaves an organization, not only there is a loss of knowledge base, but it also pulls back the momentum obtained in the project.

It is important to recognize that in practice these hazards exist and the charter should provision for mitigating the impact of the risks of the types mentioned above. Risk analysis helps in categorizing risks as high, medium, and low depending upon the *likelihood*. The likelihood is determined by the probability of occurrence of the risk—by doing a statistical analysis. Sometimes people would have expertise to do this form of classification simply by looking at the choice of technology, the business process maturity, and skills of the developers.

In the context of the consumer electronics service upgrade to mobile connectivity, the following observations can be made:

1. The J2ME is a well understood platform → the implementation technology is simple. Therefore, that is a low-risk proposition.
2. The business rules are simple and well understood → Need to use different vehicle when the mobile phones are used. The user interface needs to be reworked at the database end. Therefore, it could be a medium risk.
3. Developer is generally good notwithstanding the prevailing attrition rates. So, among the six program developers—one should expect to lose at least one person or even two. The table gives the risk mitigation plan in Table 3.3. If one person from the technical staff leaves, then the risk to the project may still be quite low. If two people leave, then the project may have to be extended minimally by a month. The probability of three developer programmers leaving is very unlikely.
4. The user interface may not meet the expected usability level. This is always a possibility for the type of project we are considering here. However, since the technology used is J2ME and is relatively well understood the assessment of risk would be low. This is because it is relatively easy to develop alternate user interfaces easily → therefore it is a case of low risk.
5. The database interface for receiving structured SMS and sending SMS alerts. This is now a well understood communication mechanism → therefore, this too is in low-risk category.

There could be other considerations in risk—for instance, what happens if the mobile service provider selection gets delayed or is not able to offer the required level of service—this may happen if the provider does not have reach to some areas where the consumer electronics services are currently available. However, overall the project is at the worst has medium risk.

An essential part of the Cloud-based software development is to understand the concept of security and risk mitigations [1]. For products that operate within the four walls of an office and all transactions are conducted behind a firewall, the risks are minimal but not zero. An example is of Edward Snowden [15], who walked out with a USB with highly classified information from the National Security Agency (NSA) in 2013. Such insider attacks are hard to prevent but stakes become even higher when the information needs to travel over the public Internet, and all or part of the product may reside on a public Cloud. Information security risks are generally classified by the acronym CIA, which expands to Confidentiality, Integrity, and Availability. Let us look at each in the context of a healthcare App that collects data from hospitals and stores, it in a Cloud for later access by the authorized users.

Table 3.3 The risk mitigation plan

Number of developers leaving	Mitigation plan	Risk level
0	Not required	No risk
1	Extend the delivery date by 15 days	Low risk
2	Extend the delivery date by 1 month at least	Medium risk

1. **Confidentiality:** Healthcare data contain private information and in most countries is protected by law, e.g., HIPPA [16] and GDPR [17]. The Cloud-based product needs to strongly encrypt this data during transmission, in the storage and at runtime access in the application.
2. **Integrity:** Even when the data may be strongly encrypted, if a malicious person is able to access it, then that person can alter it rendering the data unusable. This may cause inconvenience or even a grave danger to the patients. Thus, the application data will need to be stored in a tamper proof manner. For instance, it may be blockchained.
3. **Availability:** After the data is securely protected, it needs to be available 24 × 7 to the authorized healthcare professionals. This may require that all act of humans or nature that can compromise a datacenter's uptime should be countered by having a secure copy of data in a different location for ready access.

During the risk analysis phase of a project, managers and developers review the use cases and deployment scenarios. The goal is to check for risk items in each of the above classifications. Then these are enumerated, and solutions are proposed. These are to be implemented in the architecture (e.g., additional bits for the error correcting codes (ECC)), design (redundant datasets), and deployment (e.g., a standby server and hot-swap capability) through the product's life cycle.

Microsoft Project (and other tools of the ilk) provides detailed risk analysis using techniques like Monte Carlo simulation. Such tools help in evaluating alternate strategies for risk mitigation. Often a slack or buffer is built in the project schedule to take care of uncertainties.

3.3.5.3 Quality Assurance (QA) Plan

Ideally quality assurance plan begins with the requirements analysis phase. In some methodologies like test driven development (TDD), it is recommended that the test plan be developed well before writing any code. In Agile methodology of development, quality assurance plan, in fact, is integrated through the life cycle of the project. So, the QA plan depends upon methodology accepted within the organization.

We shall, however, dwell here on the merits of QA plan and how it may show up within the project plan. The QA is central to a SW project even while PMBOK reckons projects to be a *onetime* activity meaning that it is more often not repeated than repeated. One however has to reckon with the fact a SW project delivers something that is either directly useful to the end user or it gets embedded into something that is used by end users and may even be reused or enhanced. So, a certain minimal standard must be met. The SW quality standards [18] have been defined by the Institution of Electrical and Electronics Engineers (IEEE), International standards organization (ISO), American National Standards Institute (ANSI), National Institute of science and Technology (NIST), and Software engineering institute (SEI) located in Carnegie Mellon University through its maturity models. The maturity models called Capability Maturity Model (CMM) define five distinct levels of capabilities of an organization to deliver SW. QA protocol is mandatory for organization with CMM level 2 or higher. We shall discuss these issues in a later chapter. Right now, let us briefly describe what a QA plan entails.

The ambit QA plan should cover the following:

- **Requirements analysis phase:** The requirements are analyzed to identify QA parameters.
- **Design phase:** In case it is a standalone (like a function call), then the test plan is just for it. In case it a SW that fits in some environment, then test plan is to ensure correct interface or communications.
- **Implementation phases:** During the implementation phase, a unit test plan for each module is developed. This could also include testing out third-party components.
- **Integration phase:** During the system integration phase, every application program interface may need a well identified test plan.
- **Delivery phase**: During this phase, the real user's needs are fully covered by subjecting the systems to test in the client's operating environment. For instance, it may include QA plans for multiple OS environments or frameworks like .NET or Ruby-on-rails.
- **Post delivery phase:** During this phase, on-site support and maintenance are offered.

Like every other aspect of project planning, the QA regime needs clear schedule and documentation. It is important that the charter identifies the test and quality guarantees for every *milestone* (recall a milestone is a deliverable). In addition, the QA plan would include configuration management and change control test regime, i.e., the charter ensures that quality issues are raised and reported for each milestone in each configuration.

3.3.6 Tracking, Reporting, and Communication Management

Some projects fail simply for want of proper tracking or when the internal reporting leaves much to be desired. Note that these are completely avoidable reasons for failure. SW development projects must have a well spelt out communications management strategy for successful delivery.

To illustrate this, let us revisit the example in Fig. 3.5. We notice that 3 months into the project we would need someone to work with the mobile protocol suites for structured SMS connectivity. If it takes 1 month to recruit and have a person in place, then the timeline suggests that the recruitment should be initiated some time not later than 2 months into the project. If this project is well tracked, then the team lead, knowing his needs, would initiate recruitment process in the second month through an internal communication with the project manager. Failure to do so could lead to delayed recruitment. It could also mean that the entire project goes into wait mode with the rest of the project team not able to make much progress due to the absence of the protocol expert.

In general, the communications management is devised keeping the following in mind:

- **The stakeholder(s):** For every activity or deliverable, the project identifies to (from) whom the communication is required. The stakeholder is any person with a defined role in the project. For example, it may be an end customer, or a company board member who championed the project or the test team leader. Note that all mile stones achieved or missed need to be communicated to the concerned stakeholders. A timely communication goes a long way to support risk management.
- **Reporting timelines:** The timeline guides when the report needs to be generated. A monitoring program would track the progress and current status to trigger the communication. There are SW monitoring and reporting tools available that could help drive this activity. The tools would generate alerts to remind to raise reports timely.
- **The metrics:** The metric used in the measurement help in generating report. The metric gives an idea of the extent of effort that has gone into the project and how much more needs to be done. It helps in identifying potential points of concern which in turn helps to prioritize activities. These measurements provide rationale for revisiting the schedule.
- **The format:** For most reports, the project may specify format for internal communications for project monitoring and reporting.
- **Authorization:** Almost always documents are prepared by a group. However, internal or external release of any document would need an authorization. This assumes much greater importance whenever documents are to be communicated to the end customers or project management board members.

In addition to the above, other considerations in communications management would include the following:

- The protocol of for providing guide lines and directives to the team
- Communication between collaborating teams
- Procedure for escalation of issues and their resolution

Almost for every kind of communication, there would be some standard form. With the prevalence of electronic communications, tools are used to track and keep the communications trail with time-stamping for both formal and informal communications. Note that the time-stamping can help in also getting the correct picture to even establish the rationale for arriving at a certain decision.

More recent methods of communication include Microsoft Sharepoint, Wiki, blogs, and networking. Such communications are very useful in distributed collaborative projects. As of this book's writing, new tools are available [19].

3.3.7 Project Closure

The project closure is a major milestone. It is the time when the responsibilities get transferred. The receiver is now the owner of the developed artifact. The knowledge transfer is more effective if the receiver has been involved during the development or at least been involved with the review process. Otherwise, the project closure can be enforced only after a period of training and final handover. In our example of consumer electronics services, we had incorporated an end-user training and also a cut-over period to go live with the enhanced level of service offering.

Every organization is likely to have its own protocol for SW project closure. In fact, there are standard templates and tools available to generate a closure document using Docusign[8] and Arena.[9] From our standpoint what we need to understand is what should minimally go into a project closure document. The following are some of the critically important considerations in effecting a project's closure:

- **Documentation for the project deliverables including key takeaways:** This is the most important deliverable on project closure. In the context of our example, for consumer electronics services, it would include following.
  - The work flow and architectural details of the SW.
  - The details of any APIs and the way these APIs would invoke the services:
  - A user manual and a maintenance manual. The maintenance manual should contain the list of expected messages to help in maintenance of the system.
  - The test programs and their details. The tests may be conducted to ping the health of the system.
  - Documentation to support the fault tracking and services, such as Jira.[10]
- **Closure of contract agreements**: The following kinds of contracts may require closure.
  - The agreement with the client who needed the artifact in the first place.
  - All the agreements that may have been entered with other subcontractors who may have been engaged for specific development.
  - This may also include proper contract closure with the people who may have been temporarily engaged to work on the project.
- **Cleanup:** There is considerable holding up of the resources when a project is underway. The cleanup may require the following:
  - Releasing the resources that were being blocked by the project. This may include people, facilities like cabins, and equipment like the servers.

[8] https://www.docusign.com/ is used to create immutable records for project agreements.

[9] https://www.arenasolutions.com/solutions/document-management/ is used to store project records for audits.

[10] https://www.atlassian.com/software/jira is a project issue tracking system.

- There would be a cleanup so that no confidential information is left on the machines that are now made available to other projects. Outsourced projects often require dedicated equipment to be assigned for security reasons.

- **Risks and perceptions:** Every project leads to some learning. It helps to make observations in the closure report about what has been learned. This may cover both management and technical aspects. The technical aspects may cover the architecture or how specific design issues were handled. The management aspects may include human management, risk assessment, and estimation methodology—all of which helps in planning of future projects. The following points may be covered in the closure report:
  - Often it is important to draw upon the experience on effort estimation. Recall we earlier alluded to experts giving their estimates of the likely effort that will be required.
  - The distilled experience has no substitute in human management.
  - The nature of risks and how the management mitigated the risks. The decision process is useful and instructive.
  - The modes of communication used in the project and how effective these procedures were. This assumes importance in determining how the motivational levels may be sustained as also how escalations may be handled.

- **The continuing support and feedback:** This part provides a window of opportunity to the receiver to establish communication with the provider or developer. In fact, it benefits both. The project implementers would get important feedback on how well they performed on the end-user requirements. Also, the receiver benefits from the developer experience.

As is obvious the project closure report is the last report to be generated. It may be very enriching and illuminating.

3.4 Cloud Support for IoT

Due to the rate and volume of IoT data being generated, it is nearly impossible for humans to draw meaningful inferences. IoT is reviving expert systems and artificial intelligence (AI), aided by unprecedented computing power and self-learning systems that improve with larger volumes of incoming data. Some of the use cases for IoT-based machine learning are shown in Fig. 3.6. This figure depicts a smart home, where temperature control is based on its occupants' absence or presence. Furthermore, different parts of the smart home can be heated or cooled at different levels, instead of a single setting for the whole home.

At a basic level, we can add intelligence to remote devices if processing elements and storage are used to locally collect data for rule-based control decisions. Such smart devices offer desired functionality and operate in an energy-efficient manner with minimal computing power and memory while connected to a mobile app with

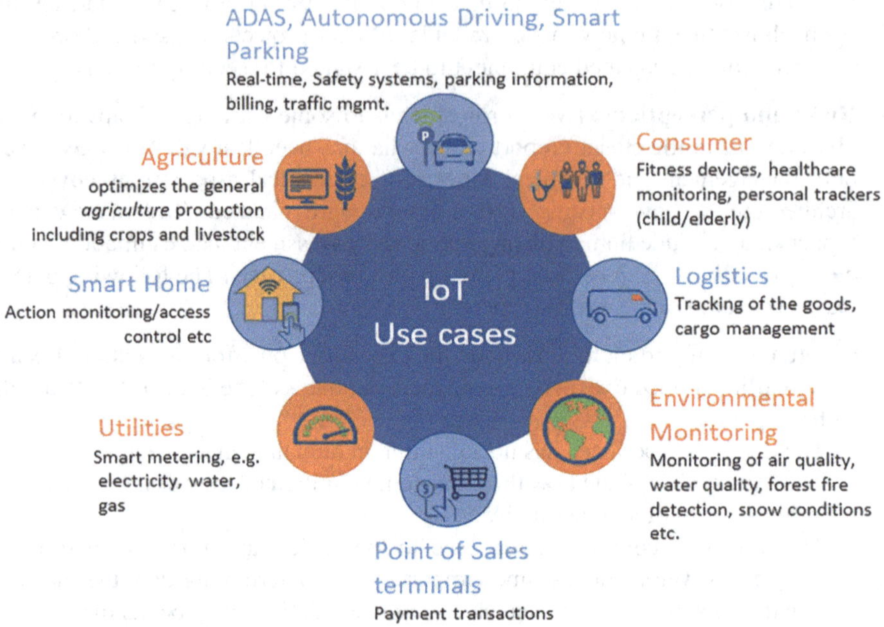

Fig. 3.6 IoT use case that needs both local and Cloud computing [20]

the Cloud on the back end. Addition of Cloud significantly enhances the scalability [1] of IoT devices as compared to standalone devices.

A Cloud service where intelligence extends beyond a datacenter to the edge-based sensors is known as fog computing [21]. What distinguishes fog computing is the way data gathering is delegated remotely and additionally the processing is localized. In this model, sensors and other connected devices such as cameras send data to a nearby edge computing device, which has processing power to analyze this data, make some local decisions, and then send the results to the Cloud. Business Intelligence (BI) forecasts that over 17 billion IoT devices owned by enterprises and governments will use fog computing in 2025, up from 570 million devices in 2015, as shown in Fig. 3.7.

An example of fog computing comes from the mining industry [22], where drilling equipment operates in harsh conditions, with autonomous trucks and trains and tunneling and boring machines moving at high speeds. To ensure the safety of workers and increase their productivity, decisions need to be made locally. Even though mining equipment can generate terabytes of data/hour during normal operation, there may not be a reliable connection to backend Cloud given hundreds of feet of underground operation, as in a coal mine. This is where fog computing can help by processing the data locally, making appropriate decisions, and as shown in Fig. 3.8, send only small uploads to the backend Cloud every few hours or at the end of each day.

Note that different legal entities may own the local sensors, edge gateway, and backend Cloud. Therefore, this may create problems with ensuring accountability

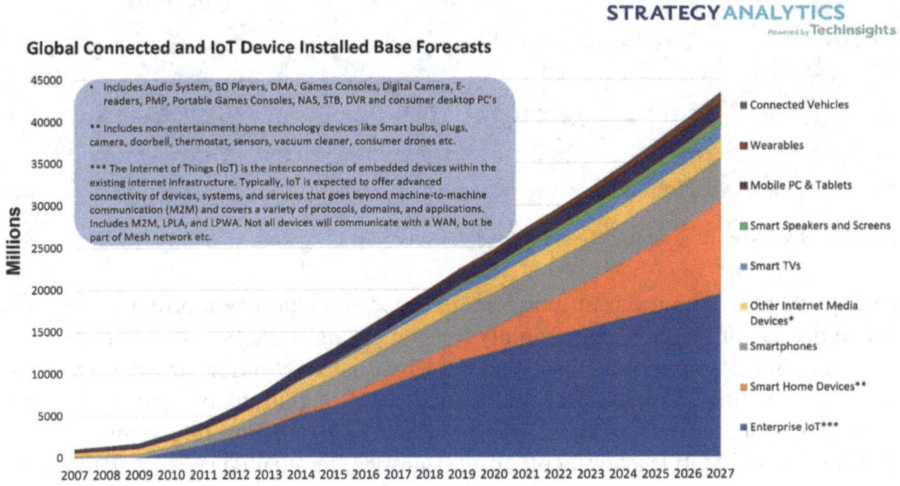

Fig. 3.7 Growth of IoT and Cloud together [21]

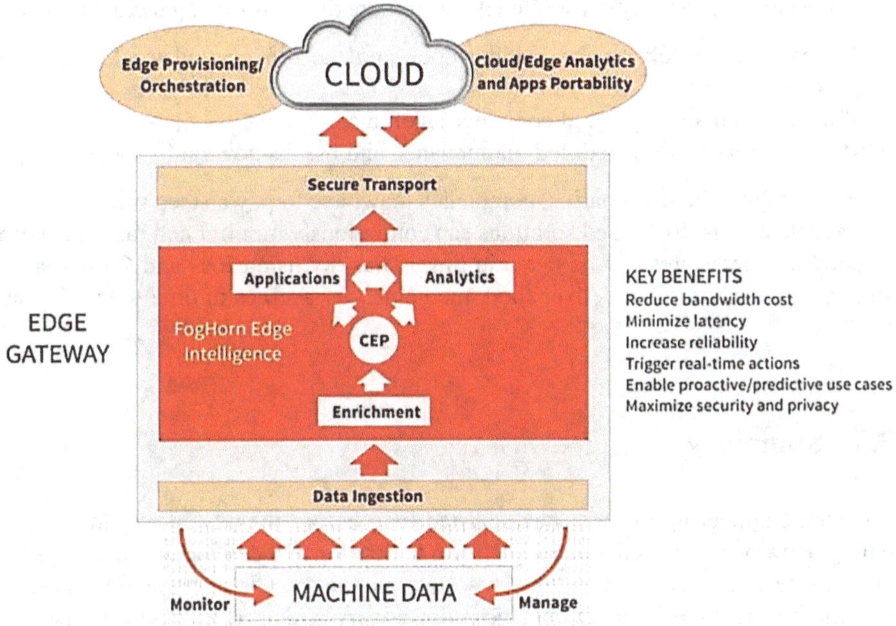

Fig. 3.8 An example of fog computing to support real-time decisions [22]

and security. If something goes wrong, e.g., in the event of a mining accident, then finger pointing will begin with hard to assign liabilities. This is where companies offering end-to-end services will have an advantage in edge computing deployment resulting in a business success. In the long run, as industry evolves and standards

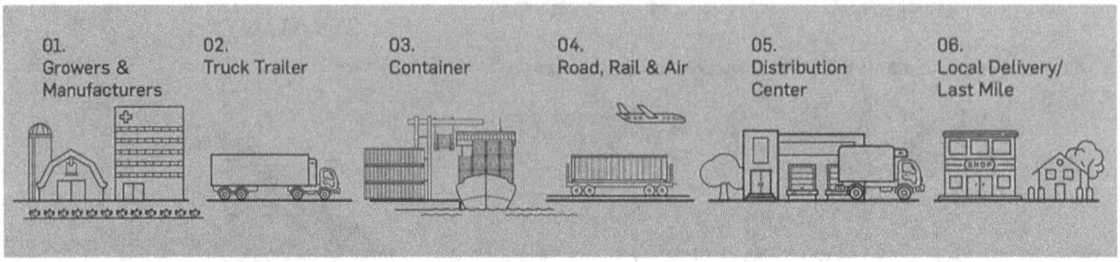

Fig. 3.9 Six stages of a supply chain for food delivery [23]

emerge, there will be room for horizontal service and equipment providers to excel at competitive price points, but initially fog computing is likely to be a vertical play.

A real-life example of using IoT in the Cloud is AWS-based services and solutions. These enable connectivity and management of billions of devices, for data collection, storage, and analytics. AWS and carrier are collaborating to innovate and develop Lynx [23]. It is a digital platform to unify a fragmented cold chain to reduce food spoilage. This is done by providing end-to-end visibility with sensors and improving energy efficiency throughout the 6 stages of refrigerated storage and transportation, as shown in Fig. 3.9.

For example, food may get spoiled if one or more of the following incidents occur:

1. If fresh food is left on the loading dock and the refrigerated truck is late in picking up.
2. If the door on a refrigerated container is left ajar.
3. If a truck misses its scheduled maintenance and breaks down mid-route.

In any of the above scenarios, companies may have to pay a steep price for inefficient cold chains. IoT-based solutions can continuously monitor and raise an alarm so corrective measures can be taken in time. Thus, by using IoT- and Cloud-based solutions, companies can deliver food, medicine, or vaccines in time with minimal wastage.

3.5 Summary

Software engineering projects have existed long before the advent of Cloud computing. However, using Cloud as a platform to develop and deliver application products has been a game changer for the software industry. Cloud-based SW product development involves product managers, developers, and users across the globe giving rise to new complexities and opportunities. A Cloud-based product plan begins with a business case approval and ends with a win-win situation for all stakeholders. Combining IoT and Cloud opens up new opportunities to improve business efficiencies. In this chapter, we described the roles and responsibilities of players involved to deliver a successful Cloud-based project within the allocated time and financial constraints.

Appendices

Appendix 1: Business Plan Template

Every SW development project is based on a specific business plan—which reflects strategic positioning of the project. The business plan for SW development project would be somewhat different from the usual commercial business plans in certain aspects. Nonetheless it should be such that anyone looking at it should understand the rationale for taking up the project. Keeping that in view, we give one sample template along with an explanation on the type of information that goes into business plans.

The Template

1. **The title page**: It identifies the company by name, logo, and location.
2. **The table of contents**: It describes the contents spread.
3. **The executive summary:** Often the summary is delimited to be less than two pages. It should provide an overview of the SW development project. It should encourage readers to read on further (to know more about the project and its likely impact).
4. **The vision statement**: It is a short and crisp statement—often less than a page—that aligns with the organization's business goals. It may refer to some initial background that helps in understanding the stated vision. The objectives are stated clearly and precisely emphasizing what will change once the SW artifact is offered and pressed into operation.
5. **The operational plan**: It shows the road map identifying some major activities and intermediate milestones. The operational plan basically includes the schedule.
6. **Description of the nature of the project**: This may describe the SW service or product with its capability. It should clearly identify the specific value proposition for the client or end user. The section may also quantify the benefits by comparing it with existing competitive services or products.
7. **The basic business set up**: It identifies the ownership structure and the nature of operations. The legal framework within which the operation can be carried out. For example, in India all cellular service providers have specific legal obligations within the circles of their operations. It should also identify roles and responsibilities within the management and the SW project team.
8. **The business strategy**: In fact, this is where one has the description of service or product distribution strategies. It should provide details of the product licensing policy or rationale for the service charges. Sometimes there may be a provision to extend and resell or repackage the SW.
9. **Cloud cost estimates:** A choice of various Cloud vendors should be considered to decide if a private vs. public Cloud will be used [1]. This appropriate cost

model keeping in mind the product deployment locations and data growth scenarios. If necessary, a multi-Cloud strategy may be adopted to avoid single point of failures [24].

10. **The financial plans**: It unfolds the funding patterns and the budget provisions. Many times, this is refined to reflect beyond the year or on a month-on-month basis. It could also project from when the organization may start deriving business value and the quantum of the same.
11. **Appendices**: There may be relevant appendices that describe other details about the employees and the general framework of administration.

Appendix 2: Project Charter Template

Once a business plan has been approved by the management, a project charter is prepared. The charter is essentially a tactical plan and reflects the course the project would take. It not only delineates the road map but also identifies detailed project management structure with roles and responsibilities for the project activities. The charter also specifies expectations, risks, and monitoring mechanism. Below we give one sample template that captures the essence of the charter.

The Template

1. **Title and table of contents:** The first page would be a title followed by sectionalized description of the contents of the charter document.
2. **Introduction to the project**: This section may have two parts—one dealing with the overall vision and the other that offers primarily an overview of the project. In essence, it *states the objectives* that have been laid out within the bounds of the assumptions made. It is also important to *highlight all the operational constraints* here. It ought to also provide information on the composite character of the project identifying the interfaces with various stake holders. It should conclude with a statement on the benefits that are likely to accrue.
3. **Statement of scope**: This section should set the context of the project and offer technical insight. Every project is positioned in some particular context in a specific domain. For example, the domain of its definition may be like web application or services, enterprise resource planning, etc. The scope identifies the boundaries of the infrastructure needed and the way this infrastructure will help in realizing the project objectives. Finally, the section must define the specific functionality attained within the domain of its operation along with the business value proposition as envisaged in the original business plan.
4. **Organizational structure with roles and responsibilities**: This section gives a clear definition of the executive structure of the project, as follows.

    (a) **The project office structure:** This subsection gives a detailed project management structure. It includes the following.

   (i) The managers and project lead with their roles.
   (ii) The project administration (including the accounting section) for day-to-day admin support in executing the project.
   (iii) SW management including system administration, configuration management, change control, and management set up.

(b) **Monitoring and the oversight framework:** These refer to the executive committees and their terms of reference. It would include all the reviews that shall be due during the lifetime of the project.
(c) **Interfaces:** These are to deal with vendors, suppliers, or contractors who may be outsourced some part of the work.
(d) **The quality assurance:** This is to ensure that the evidence we use (to support quality claims) assures meeting quality standards. Also, adequate records are maintained to support an audit with accountability.

5. **Project plan**: The project plan has a detailed description of how the project shall be realized. It aligns itself with the scope (section—2) and covers several facets that include the following:

   (a) **Project schedule:** It contains project start and end dates.
   (b) **Human resource:** This includes plans on when and how the people would be inducted. Some changes of personnel may be required when project phase transitions, i.e., at the end of or beginning of each phase there should be smooth hand over to ensure project delivery on time. This may include specific training plans for project personnel.
   (c) **The timelines marked with milestones:** In fact, the project phases and their deliverable are identified. It may have a detailed break-up of the activities and even tasks.
   (d) **Project review dates** to help in monitoring the project's progress.
   (e) **Quality assurance:** It may delineate acceptance criterion for various stages of the project.

   Together with (c) it ensures that the project is on track to deliver the tangibles.

   (f) **Risks and risk mitigation plans:** Risks refer to identification of threats faced by a project and product. Mitigation plan is a strategy to reduce the effects of these threats.

6. **Project procedure guidelines**: The guidelines help in quick authorization and approvals required from the correct levels in management. This helps in keeping up with the project schedule. It lays out how to manage to limit escalations. For instance, error reporting and the change management should happen such that the project activity continues with minimum disruption while maintaining the team harmony.
7. **Sign off and closure procedure**: Most organizations have some kind of approval protocol to close a project. This protocol should prescribe all the approvals required before the project is handed over to the end user with an agreed signing off from the project.

Appendix 3: PERT and Critical Chain Scheduling

The Program Evaluation and Review Technique (PERT) is used to estimate time for a Project's activities with many unknown factors [25]. This technique was promoted by US Navy and has been in use now for over six decades. It uses weighted average of the Optimistic (O), Pessimistic (P), and Most Likely (M) estimates in the presence of uncertainty with individual activity estimates. PERT uses following 3 estimates:

1. **Optimistic (O) estimate:** O is the best-case and thus lowest time or cost to complete the work.
2. **Pessimistic (P) estimate:** P is the worst-case and thus longest time or cost to complete the work. This assumes all negative risks occurring with no mitigations in place.
3. **Most Likely (M) estimate:** M captures the highest likelihood of completing the work in the given duration or cost. This assumes some risks occurring.

The PERT formula computes time as (O + 4 M + P)/6, using the above estimates. The M value is given 4 times weight based on the probability theory and statistics specific beta distribution [26]. Such a distribution results in a bell-shaped curve, aka normal distribution.

In practice, PERT integrates uncertainty and makes use of assumptions that can accelerate or delay a project's progress. This gives project managers some estimates of the possible time variations for offering safe delivery dates to the customers. Some managers also like to use Critical Chain Scheduling (CCS) alongside PERT. So, next we describe CCS briefly.

Typically, a project consists of a series of tasks. The critical chain of a project lists the dependent tasks that define the lower limit of possible lead time [27]. Even if other tasks are sped up or eliminated, the completion time for a project will not shrink, because the critical chain is made up of a sequence of dependent tasks. In CCS, the output of a task forms the input to another, such that the latter cannot be started until the former is completed. Such a dependency can come from the nature of tasks, e.g., coding for an algorithm cannot commence until an algorithm is selected. Dependency may also arise from the availability of a resource. For example, a test engineer needs to finish verification of one block before she can start the other.

CCS accounts for it using a buffer, as shown in a representative Fig. 3.10. A buffer is computed and applied to ensure the success of a project in view of the uncertainties. Its aim is to protect the delivery dates from variations in the critical chain tasks. Some tasks that appear to be in series, but are not dependent on each other can be performed in parallel (with time overlap) if sufficient additional resources are available.

A "resource buffer" reflects the availability of additional resources that parallelize two tasks in a chain, such as A1 and A2, or A3 and A4. For example, this may pertain to the availability of multiple test engineers who can test these blocks in

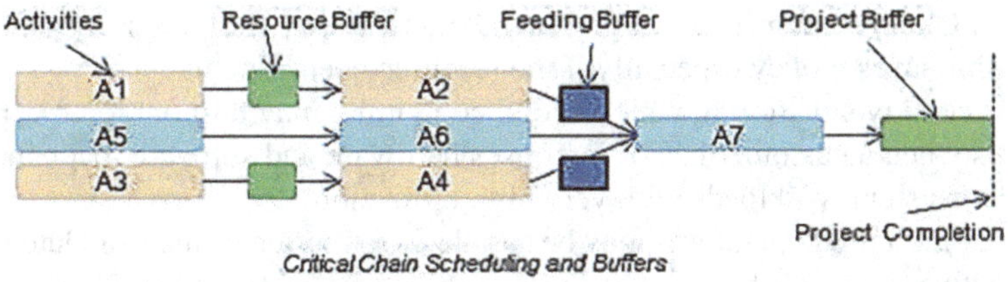

Fig. 3.10 Using buffers to ensure success of a project [27]

overlapped time frames, as no inherent task dependencies exist except of available resources. In Fig. 3.10, we have a "feeding buffer" which indicates that the start of A7 is not dependent on the completion of A2 or A4, so A7 can start as soon as A6 is done. Thus, the critical path for the example in Fig. 3.10 consists of tasks A5→A6→A7.

Often there are multiple projects running in an organization. For such cases, CCS uses a "project buffer" at the end to indicate dependence on some key resources that may be involved with other projects. A variation or delay in one of these projects may impact the completion of the current project. A *critical path* goes from the start of a project to the end [20], whereas a *critical chain* ends at the start of the buffer assigned to that project. Hence, the activity sequence in a critical path defines performance with less emphasis on the resource utilization, but focuses on the milestones and deadlines.

Points to Ponder

1. Why is Cloud computing considered cheaper than on-site dedicated resources?
2. Why must every software project start with a business case?
3. What happens to the budget of a project if it takes longer to deliver?
4. What is the role of a DevOps engineer?
5. How can the security of a Cloud-based software project be assessed?
6. How does IoT enhance Cloud-based products and services?

Answers

1. Why is Cloud computing considered cheaper than on-site dedicated resources?
   (a) Mainframe computers introduced the concept of time sharing and Cloud is one of ways to share resources among multiple users.
   (b) Cloud enables sharing of resources, so computing tasks from many different customers can use the same underlying hardware. This improves resource utilization and in turn, Cloud service providers pass some of the savings to their customers.
   (c) Cloud computing is generally cheaper than on-site resources, as dedicated hardware and software licenses and IT personnel do not need to be maintained.

(d) A Cloud customer uses a pay-as-go model to pay for the services used, and this saves money especially if the resource needs fluctuate.
(e) The only situation in which Cloud computing may turn out to be expensive is when a customer has on-premises hardware and software that can handle consistent workloads with very high utilization.
(f) Cloud-based operations may be avoided for mission critical and latency sensitive systems when cost and utilization of resources can be overlooked.

2. Why must every software project start with a business case?

   (a) In a capital driven society, money fuels all activities and is the starting point of funding a software project. A new project, or revision of an existing one, should be undertaken only if a positive ROI (Return on Investment) can be shown by assessing the need for a new work, its investment requirement, and payback indicated.
   (b) Often a make vs. buy analysis is also needed to see if a software package or product should be purchased, customized, and deployed instead of an in-house development.

3. What happens to the budget of a project if it takes longer to deliver?

   (a) Budget is a commitment to provide resources for the project. Budget is the enabler of activities. Only on being budgeted an activity be initiated. It is quite possible that during the lifetime of a project, it is realized that the budget did not provide for some essential activity. In that case the budget would need to be "adjusted." In other words, the budget would be modified before the activity is permitted to go ahead.

4. What is the role of a DevOps engineer?

   (a) A DevOps (Development and Operations) engineer guides software developer to make code releases and helps with all the operational aspects including product deployments in the Cloud.
   (b) DevOps engineer also helps to build team spirit by encouraging team members to collaborate and meet each other's requirements.

5. How can the security of a Cloud-based software project be assessed?

   (a) Information security risks are generally classified by the acronym CIA, which expands to Confidentiality, Integrity, and Availability.
   (b) **Confidentiality:** The Cloud-based product needs to strongly encrypt data during transmission, in the storage and at runtime access in the application.
   (c) **Integrity:** Even though the data may be strongly encrypted, if a malicious person alters it rendering the data unusable, it may cause inconvenience or even a grave danger. Thus, the application data will need to be stored in a tamper proof manner. This may use error detection and correction codes.
   (d) **Availability:** After the data is securely protected, it needs to be available 24 × 7. This may require that all acts of humans or nature that can compromise a datacenter's up time should be countered by having a second copy of data at a different location for ready access.

(e) A security assessor needs to examine the attack surface to assess the severity of possible attacks on a Cloud-based software project, then build the threat model by considering different players and vectors using the CIA framework.
(f) Each risk needs to be enumerated and mitigated.

6. How does IoT enhance Cloud-based products and services?

(a) IoT enables a Cloud service to be enhanced using edge-based sensors.
(b) IoT-based solutions can continuously monitor and raise an alarm so corrective measures can be taken in time.
(c) IoT sensors provide real-time data and analytics from the field for better decision-making in the Cloud.
(d) An example is of traffic cameras at intersections that can detect accidents and indicate the situation to drivers on their way to avoid potential traffic jams.

References

1. Sehgal, N. K., et al. (2023). *Cloud computing with security and scalability*. Springer. https://link.springer.com/book/10.1007/978-3-031-07242-0
2. Mell, P., & Grance, T. (2009). *The NIST definition of cloud computing* (version 15 ed.). http://csrc.nist.gov/groups/SNS/Cloud-computing/Cloud-def-v15.doc
3. Armbrust, M., Fox, A., Griffith, R., Joseph, A. D., Katz, R., Konwinski, A., Lee, G., Patterson, D., Rabkin, A., Stoica, I., & Zaharia, M. (2010). A view of cloud computing. *ACM Communications, 53*, 50–58.
4. Ramgovind, S., Eloff, M. M., & Smith, E. (2010). *The management of security in cloud computing* (pp. 1–7). Information Security for South Africa (ISSA).
5. Marchekwa, J. T. *Information technology project management*. Wiley 1st edition 2003, 3rd edition 2009, Also see course slides by Prof. Marchekwa, North Illinois University.
6. https://en.wikipedia.org/wiki/Cell_(microprocessor)
7. https://www.gamespot.com/articles/ps-now-explained-what-you-need-to-know-about-sonys-cloud-gaming-service/1100-6471137/
8. https://businessmodelanalyst.com/uber-business-model/
9. Berkun, S. (2005). *The art of project management*. O'Reilly.
10. McLeod, G., & Smith, D. (1992). *Managing information technology projects*. Boyd and Fraser. http://www.projectperfect.com.au/info_define_the_scope.php
11. Stallman, A., & Greene, J. (2005). *Applied software project management*. O'Reilly.
12. https://semaphoreci.com/blog/cicd-pipeline
13. Fenton, N. E., & Pfleeger, S. L. (2004). *Software metrics*. Thompson Books.
14. Futrell, R. T., Shafer, D. F., & Shafer, L. I. (2008). *Quality software project management*. Pearson Education, 4th print.
15. https://en.wikipedia.org/wiki/Edward_Snowden
16. https://www.cdc.gov/phlp/publications/topic/hipaa.html
17. https://gdpr-info.eu/
18. Covey, S. Notes from Cosntrux.com
19. https://www.comindwork.com/
20. https://www.telecoms.com/iot/nb-iot-setting-the-pace-in-the-race-to-5g
21. https://www.linkedin.com/posts/ekta-mittal-_the-connected-and-iot-device-market-is-big-activity-6977909626168094720-u-ba

22. https://www.nanalyze.com/2016/08/fog-computing-examples/
23. https://aws.amazon.com/blogs/industries/carrier-and-aws-collaborate-to-reduce-food-spoilage-across-the-cold-chain/
24. https://www.gartner.com/smarterwithgartner/why-organizations-choose-a-multicloud-strategy
25. https://projectmanagementacademy.net/resources/blog/a-three-point-estimating-technique-pert/
26. https://www.statlect.com/probability-distributions/beta-distribution
27. https://www.tutorialspoint.com/management_concepts/critical_chain_scheduling.htm

Chapter 4
Cloud Product Development Process

4.1 Introduction

This chapter is primarily concerned with software product development processes including its management. Specifically, it emphasizes developing software for deployment in Public Clouds. We begin the chapter by explaining the need to do a Proof of Concept (PoC) to demonstrate the feasibility of an idea. This chapter describes two development processes, namely Agile and DevOps. We also explore and emphasize using software testing to ensure that the final product meets given requirements.

4.2 Proof of Concept

Before embarking on a full-fledged Cloud-based project's execution, or to plan resources for it, one needs to ascertain the feasibility of a project. The primary purpose of doing a Proof of Concept (PoC) is to gain confidence in the viability of the idea, technology, and team's ability to deliver on the project. A PoC can be seen as a stripped-down version with minimal GUI and I/O (input-output) capabilities primarily to convince the stakeholders that the desired goals are achievable. We briefly discuss Progressive Web Apps (PWA) development for developing a PoC. Later, in Chap. 9, we shall probe and compare two distinct approaches, namely the PWA and native application development, primarily detailing their pros and cons. PWA is often a preferred option for rapid prototyping for PoCs.

A PoC needs to clearly state the assumptions being made, because if these are unrealistic, then the PoC will work but the real project may not. An example is the famous whitepaper for crypto currency [1], which proposed a solution for peer-to-peer electronic cash transactions. It unleashed a whole new industry with

P. C. P. Bhatt et al., *Project Management in Cloud and IoT Applications*,
https://doi.org/10.1007/978-3-031-99427-2_4

public proof of concept, as shown in Fig. 4.1. Crypto currency transactions are both reliable and secured by a chain of digital signatures. An electronic coin was defined, and all the transactions using the coins are recorded with a chain of digital signatures. A coin owner transfers coins by digitally signing a hash of the previous transaction and the public key of the receiver. This transaction is added to the end of the chain. A payee can verify the signatures to verify the chain. The blockchain ensures that the man-in-the middle attacks are rendered ineffective even though the transactions are open (for examination). However, the ownership is not revealed and remains inaccessible.

The proposed solution eliminated the need for a trusted central authority, or mint for monetary transactions. This was accomplished by a public or shared ledger where each transaction is time-stamped and recorded as a hash. This scheme also helps in preventing double-spend. Each timestamp includes the previous timestamp in its hash, forming a chain, aka blockchain, as shown in Fig. 4.2. Anyone can see that someone is sending an amount to someone else, but without information linking the transaction specifics. This ensures complete privacy. This white

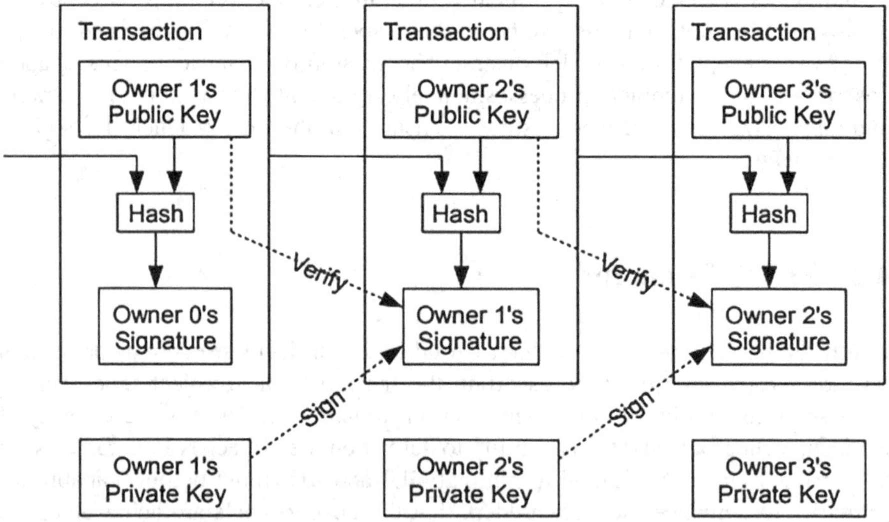

Fig. 4.1 Verification for chain of ownership in digital cash transactions [1]

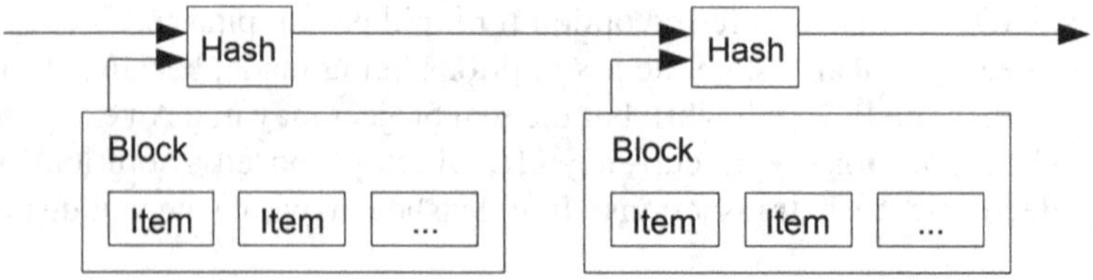

Fig. 4.2 A blockchain example [1]

paper and its prototype resulted in a system for electronic transactions without relying on any intermediary. Multiple copies of the public ledger show a record of history of transactions that quickly become computationally impractical for a hacker to alter in anyway.

Currently, many versions and types of digital currencies have appeared on the horizon relying on the hashed digital signatures with timestamps in a blockchain that is publicly available. None of these would have been possible without the first whitepaper and a Bitcoin prototype built in the public domain.

Nowadays, it is a widely accepted practice that any complex software project should undergo the rigor of validation of the idea by first prototyping, followed by iterative improvements to achieve the final goals.

4.3 Traditional Software Development Methods

In the area of software development, Waterfall model [2] was the first process model to be introduced. It is also called a linear-sequential life cycle model. It is very simple to understand and use. As shown in Fig. 4.3, it has multiple phases, and each phase must be completed before the next phase begins. The phases in this model do not overlap in time, and there is no provision to go back. If some requirements change midway, their implementation must wait till the next project cycle or the current project may be abandoned to start all over again. Note that the classical Waterfall model has no feedback loops. In fact, it mimics water falling down the steps in real life. However, in a modified version, feedback from one phase is carried back to the previous phase. The feedback is shown as dotted lines in Fig. 4.4. Once the feedback is incorporated, the Waterfall process resumes. However, there may be a specific case when one step backwards may result in subsequent feedback steps. Imagine an example if during deployment, a bug is found, then during test analysis, a gap is identified in the coding stage, which traces back to a design issue that may result in alteration of the original specifications, effectively restarting the Waterfall process.

Given this obvious disadvantage, as requirements do change in the real world during the duration of a project, very few software projects nowadays follow the Waterfall model. Before moving to other methods in vogue, it would be instructive to visit the advantages and disadvantages of the Waterfall model [2]:

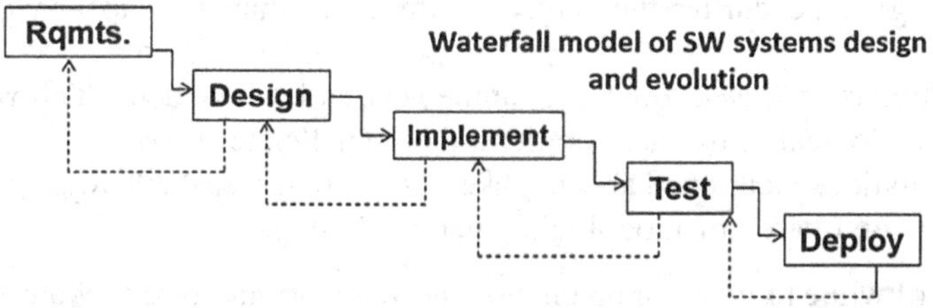

Fig. 4.3 A pictorial representation of Waterfall development model [2]

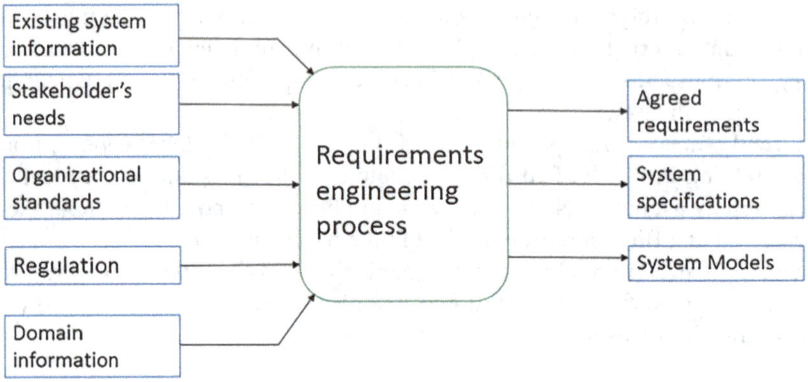

Fig. 4.4 Inputs for developing requirements and output from the process [4]

Advantages

1. Simple and easy to understand and use.
2. Easy to manage due to the rigidity of the model. Each phase has specific deliverables and a cleanly defined review process.
3. Phases are processed and completed one at a time.
4. Works well for projects where requirements are very well understood.
5. Works well for projects where the execution team has prior experience.
6. Clearly defined stages.
7. Well understood milestones.
8. Easy to arrange tasks.
9. Process and results are well documented.

Disadvantages

1. No working software is produced until late in the project life cycle.
2. Not a good model for complex and object-oriented projects.
3. Poor model for long and ongoing projects.
4. Not suitable for the projects where requirements are at a moderate to high risk of changing. So, the risk and uncertainty are high with this process model.
5. It is difficult to measure progress within stages.
6. Cannot accommodate changing requirements.
7. Adjusting scope during the project's life cycle may lead to abandoning the project.
8. Integration is done as a "big bang" at the very end, which doesn't allow identifying technological or business bottlenecks or challenges early.
9. The limitations mentioned above make Waterfall non-scalable for new technologies or Cloud-based methodologies such as DevOps.

It is worthwhile to mention one major success story in favor of Waterfall model. In 1972 (Agile was not much in vogue then), five employees of IBM in Germany left to form a company called SAP [3]. All that they initially offered was a more usable front-end to the Databases running on IBM machines. They began by supporting some existing IBM clients for Accounting, Human Resource, and Logistics. Later, they acquired their own servers and began operating as an IT service company. All this was possible due to the deep knowledge and understanding of each stage of the process. They progressively moved to serve other segments within a client's company and later across domains such as aerospace, mining, and retail. Subsequently, they graduated to e-commerce operations for faster resolutions.

The example above clearly indicates following attributes of a successful Waterfall project:

1. Requirements are very well documented, clear, and fixed.
2. Product definition is stable.
3. Technology and techniques are very well understood, stable, and widely accepted.
4. There are no ambiguous requirements.

In later sections of this chapter, we discuss iterative development methodologies to demonstrate how one incrementally builds up the final product and how this helps in gaining stakeholders' confidence in the process. First let us discuss how the requirements for a major SW product development activity may be generated.

4.4 Requirements Gathering

Product requirements are defined in the early stages of a project's development. A cleaner statement of the requirements can lead to implementable design specifications. The requirements may be arrived at in multiple ways. They may also be stated in multiple ways. The processes leading to requirements are broadly covered under the umbrella term "Requirements Engineering." The objective is to state "what" is to be developed without elaborating the "how" part. The "how" part usually covers platforms, detailed software design, coding and testing, and the types of software tools to be used besides the aspects relating to scheduling of tasks and activities. So, the requirements engineering is relatively a very small part of the product development activity. However, errors in requirements specifications get amplified in later stages with catastrophic consequences. The requirements gathering is shown in Fig. 4.4. It is a very crucial and important stage in the software development [4].

A good requirement ought to be complete, consistent, and relevant. These three properties are defined below:

1. **Complete**:

   - Complete means comprehensive coverage of every aspect that needs to be realized for the artifact.

     – Requirements stated with clarity, i.e., not subject to different interpretations.
     – The documentation shall be an unambiguous source of reference.

2. **Consistent**:
   - There are no built-in contradictions in the requirements.
3. **Relevant**:
   - The requirements have a context. They capture every relevant facet of the application and the likely operational environment.

The inputs to develop a product's requirements come from:

1. **Domain information**
   - Like banking, process control, materials management, etc.
2. **Internal and regulatory standards**
   - Like HIPPA for medical care, Basel standards in Banking, ISO in process control, NW protocol and traffic management standards, and digital intellectual property rights in media management.
3. **Organizational standards**
   - Like "Best Practices" and policies followed within the organization.
     - Applies to financial policies and engineering procedures.
4. **Existing internal system** (which may need to be automated)
   - The current system defines the workflow which is replicated with greater efficacy, better access control for security, and coordination within the set up.
5. **The stakeholders' needs**
   - The most important stakeholder for a Cloud-based software system is the user. Clearly, every software solution has some context of use. For example, arguably worldwide the single most frequently used software would be MS Office. Microsoft understood it well and offered a flexible enough suite which could integrate with other applications. MS office offers a suite of interoperable software. For example, one may create a figure using PowerPoint and transfer its image to MS word.

Cloud-based product requirements are often classified into following four categories:

1. **Functional Requirements**
   - Input output characteristics.
     - E-commerce sites offer a basket and generate order id with details.
     - Debit card multi-factor authentication (card, PIN and OTP, etc.)
   - Functional behavior
     - May include access control privileges.
     - Search and query capabilities.
     - System state information management.

2. **Non-functional requirements**
   - Usability considerations.
     - Number of clicks needed to get information.
   - Printing and formatting considerations.
     - Payroll printing (cheques, memos) on stationery with tear-off options.
   - Performance characteristics including response times.
     - This is critical for databases, real-time operations, and embedded system operations.
     - Time taken to generate and communicate One-Time Passwords (OTP).
   - Interoperability considerations.
     - Web browser compatibility, rendering considerations.
     - NEFT or IFSC or Funds Routing codes for funds transfer.
   - Reliability
     - Mean Time Between Failures (MTBF).
     - Mean Time to Recover (MTTR).
3. **Scalability and capacity**
   - Horizontal (replicating) capabilities to enhance capacity.
   - Vertical (enhancements within the system).
   - Number of concurrent users.
   - Number of transactions.
4. **Domain requirements**
   - Domain-specific constraints and standards.
     - In banking compliance to Basel norms.
     - ISO 9000 norms.
     - Digital rights.
     - Safety norms for software (as in environments that have radiations).
   - **Security standards for the domain**
     - Multi-factor authentication.
     - Encryption standards.

The notion that requirements be complete, consistent, and relevant implies that leaving or adding something risks:

- **Worst case:** Losing some aspects of the desired or critical system's behavior requirement.
- **Best case:** Occasional ambiguity in system behavior.

Once the requirements gathering is completed, two distinct development methodologies are available, namely Agile and DevOps. We shall review them in the following sections.

4.5 Agile Development

A software development project invariably aims at either creating a new product or a substantive improvement over a current offering. Also, the implementation is almost always carried out by a group. So, the challenge is that the group should remain motivated while sustaining a certain pace of development. In the case of a new product, sometimes a new framework needs to be formulated. This may be because the old frameworks may be of limited or no use to achieve substantive improvement. Therefore, one must be prepared for several missteps as the project evolves. Additionally, project development effort must adapt to evolving requirements with mid-course revisions or corrections. The objective then would be to keep the team's enthusiasm and manage group dynamics. This is achieved through regular brainstorming and continuous evaluation. *Regular brain storming and continuous evaluations are the corner stones of the Agile program development methodology.*

Before we delve into defining and understanding Agile development, it is instructive to learn about iterative development process. It starts with a small set of requirements and its implementation. Then it iteratively enhances the scope with a few additional requirements—each time creating a newer version until the complete system is implemented and ready to be deployed. In other words, an iterative life cycle mode does not even attempt to start with a full specification of requirements, but for a part of the software. This is then used to identify additional requirements. This process is iteratively repeated, until the full system is ready.

Another variation of this is an incremental model, where complete requirements are known and divided into various builds. In every iteration, the development team picks up some additional requirements and goes through their design, implementation, and testing phases. Thus, each subsequent release of the module adds functions to the previous release. This process goes on till the complete system is ready as per the full set of requirements.

Let us begin by answering one question: When did Agile process formally come on the horizon? Kent Beck, Ward Cunningham, Arie van Bennekum, Alistair Cockburn, and some others, met for what is now famously referred to as "Snowbird meeting" in Utah (USA) [5] in 2001 and formally proposed Agile process to replace the Waterfall model for software development.

Agile model [4] is a combination of the iterative and incremental processes, with a focus on customer satisfaction by rapid delivery of working software products. It breaks the product into small incremental builds. Each iteration of Agile development typically involves the following steps:

1. Planning.
2. Requirements analysis.
3. Design.
4. Coding.
5. Unit testing.
6. Acceptance testing.

As depicted in Fig. 4.5, at the end of each iteration, a progress on the working product is shown and demonstrated to the stakeholders and customers. This is used to seek their feedback, refine requirements, and make measurable progress towards the final project goals.

The main point in Agile context is that each iteration is expected to realize a small goal. The goal at the end of each stage requires that a certain small set of additional features are completed or at least demonstrated. The art of being able to do so distinguishes Agile process from the usual iterative procedure of software design.

The key characteristics of the Agile process include:

1. Brainstorming.
2. Stories and scores.
3. Time boxed operations (the sprint) for realizing smaller goals with working features.
4. Continuous testing and evaluation which may require revisiting the goals.

Each iteration in Agile process results in releasing software with some additional features. Note that time boxes set the pace for sprint of the Agile process. Figure 4.6 shows the time boxed operation for the iterative process shown in Fig. 4.5. As stated earlier, each iteration in Agile process results in release of software with some additional features. Each time box takes nearly equal time and effort to advance the project by one stage. As for the number of stages n, a broad-based consensus is arrived at through discussions among all the stakeholders.

The stake holders include the following:

- End user or client who knows and understands the domain of use and workflow.
- SW architect, design team, and test team leaders from implementation team.

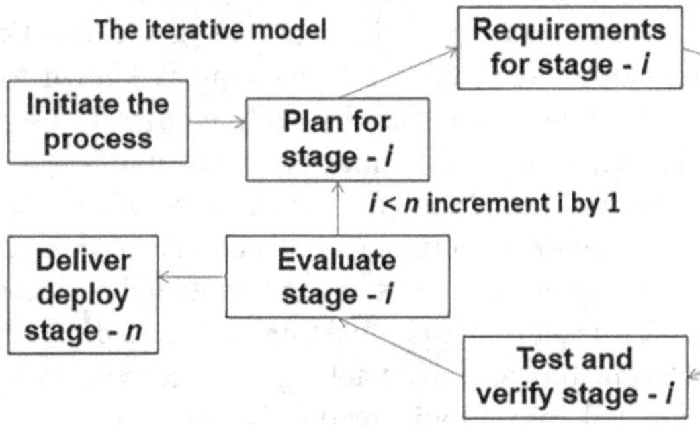

Fig. 4.5 An illustration of the Agile model [2, 4]

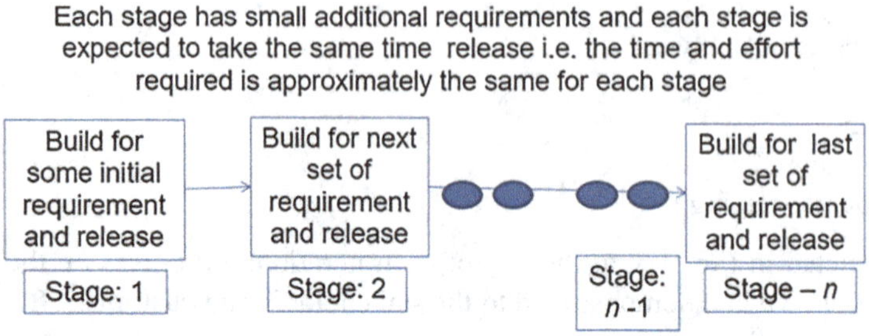

Fig. 4.6 Time boxed operations

- Market research folks who keep a tab on market acceptability and the emerging trends.
- The business associates along with possibly company advisors and mentors.

Basically, all who matter have a say. This is the Agile mechanism to resolve conflicts.

Agile uses a metric called RTF (Realized-Tested-Features) on a continuing basis. The periodic evaluations using RTF are aimed at deriving customer satisfaction and acceptance [4]. As for the management of Agile process in practice, a methodology called "Scrum" (derived from the game of Rugby) is used. *As a notion, Scrum suggests inching forward as a group.* The basic premise of Scrum project management methodology [4] requires scheduling following kinds of meetings. Usually, these meetings are directed by a Scrum Master, who may also be a CTO or a Project Director.

- **Sprint planning meeting:** This is a major meeting, and its periodicity matches the time box, i.e., its objective is to plan immediately following release and to prepare for the next one. Usually, it is expected that the team delivers a certain set of new features at the end of each time box. So, the planning meeting focuses on the next set of deliverables. The planning process must identify the resources required and the strategy that would ensure that goals are more likely to be met.
- **Daily Scrum meeting:** It's the first thing the team engages in on daily basis. This is usually a stand-up meeting of a very short duration—no more than a quarter of an hour. Its objective is to review and plan the activities of the day. For a geographically dispersed Cloud-based project, this meeting is known to resolve problems that require dovetailing of one activity with another. It also helps to sync up team members located in different time zones, i.e., mutually dependent individuals (or groups) determine what they need from some other individual or group. This helps in prioritizing on a continuous basis and keeps the momentum going. These meetings also iron out the kinks through mutual discussion.
- **Sprint review meeting:** This is an important meeting as it has a direct impact on the sprint velocity at each time box and also backlog management. Two key features in a review are directed by periodic feedback and retrospection by all concerned.

One major aspect of the Agile process is the practice of extensive testing at every stage. The test plans are prepared in advance for each unit and its integration as the system implementation progresses. Agile methodology finds greater levels of acceptance across the spectrum of stakeholders including sponsors.

Agile process has some limitations, as it depends heavily on continuous user interactions. If users are not available at end of the iterations to give feedback, then a project may potentially drift in a different direction. Often the responsibilities for requirements are spread across the Agile team. A more detailed discussion of Agile is beyond the scope of this book but can be found here [4].

Later in this book, we shall review the case study of a real-life Cloud application developed by a startup, with teams located in multiple locations in California, the USA and India, with users spread across the USA. The product development was started by first identifying an unmet user need, then developing a PoC. It was used for clinical studies of the results which were uploaded to Cloud. Based on the users' feedback, then the next version of the Cloud-based application was developed and productized for revenue. Now the application is in its fourth generation with new features being added and more customers using it with thousands of real patients' data in the Cloud.

In the next section, we will discuss the development process called DevOps.

4.6 DevOps

The short form DevOps comes from two words: *development* and *operations*. The objective here is to recommend a collaborative approach primarily between two teams—the development team and an operations team [4]. When an enterprise seeks a new solution or a major overhaul, the traditional approach follows two steps:

1. The development team engages in activity sequence: code → build → test → release. The released SW is offered to the operations team.
2. The operations team follows it up as: deploy → operate → monitor → plan.

Clearly, there are two silos of activity. Their isolation usually results in many missteps before an enterprise gets its scheme operational as the two teams may not synch. This often results in delays. Just as the Agile methodology, emphasizes a degree of communication between stake holders, the DevOps encourages the two teams to work collaboratively with feedback as shown in Fig. 4.7. In fact, often some degree of cooperative competition evolves between these two teams to meet each others' expectations and schedules.

As the loop in Fig. 4.7 shows, a DevOps team can automate the build, test, and deploy phases of a release process every time there is a code change. It enables a product team to deliver new features and updates rapidly and reliably. This is also known as the Continuous Integration and Continuous Delivery (CICD) pipe [4].

The feedback mechanism ensures that the development code is frozen only when the operational team figures out that their concerns have been adequately addressed.

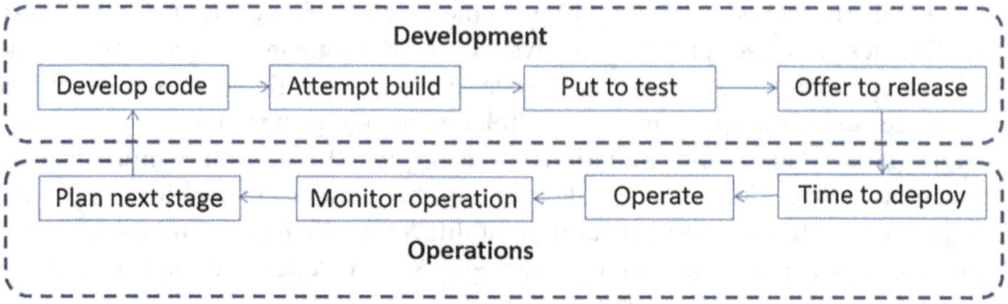

Fig. 4.7 The DevOps—a new way to develop enterprise and Cloud solutions [4]

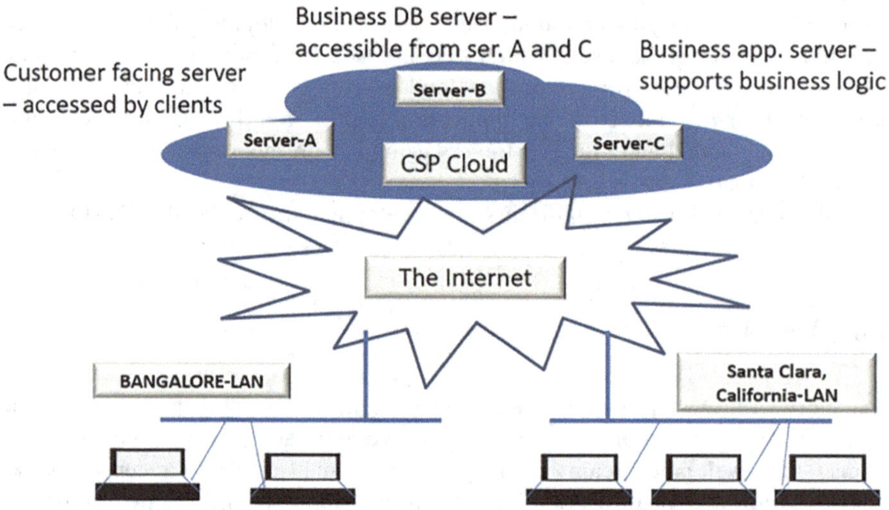

Fig. 4.8 Agile team of developers in India working with DevOps team in USA [4]

It has been found that the DevOps offer benefits in terms of better version control, configuration management with more stable releases and predictability.

4.7 Cloud Software Projects

In the Cloud context, DevOps help rolling out new releases with evolving and emerging specifications of the network and server configurations [6]. Thus, for a geographically distributed Cloud-based product development, it is perfectly logical to have development team follow the Agile process in one country, while the operations team located in another country follows the DevOps to make regular releases to customers dispersed across the world. This is shown in Fig. 4.8. The teams in India and the USA must create a synergistic operating environment.

A Cloud software project will also differ depending on whether it is using compute resources located in IaaS, PaaS, or SaaS framework. Note that IaaS (Infrastructure as a Service) uses bare bone services such as computing and storage in the Cloud. Thus, the product team will need the services of an IT manager to manage the system configuration and resources in the Cloud. A fine grain specification of resources allows for more cost control but needs frequent interventions to ensure the uptime and availability of the services.

Whereas, PaaS (Platform as a Service) usage will need a DevOps engineer to decide on the specification of higher-level Cloud services such as a load balancer to ensure needed performance and desired datacenter locations of on-demand servers to minimize latency, but extra budget needs to be planned as an incorrect specification will result in higher-than-expected Cloud bills during the product development process. Such surprises can be minimized by close tracking (daily or hourly) of Cloud logs and bills being accumulated.

Lastly, a software product using SaaS (Software as a Service) is further removed from the underlying hardware and infrastructural services, solely depending on the high-end service providers. An example is specialized AI/ML- (Artificial Intelligence and Machine Learning) based services provided by the Cloud service providers, e.g., Sagemaker by Amazon's AWS, Tensor flow from Google's GCP, and ChatGPT from Microsoft's Azure. The users are charged based on time used, and costs are usually higher than that for the IaaS and PaaS users, but SaaS minimizes system administration concerns.

4.8 Testing Cloud-Based Products

In the decade of 1950s, commercial computer systems emerged and programming came into being a profession. Professional programmers were required to develop programs to support every conceivable application. One may say the software testing took roots right then. The main concern at that time was to find and get rid of programming errors. This activity was appropriately termed as debugging. The errors were considered bugs. Debugging often assured that that the programs are likely to execute. However, Edsger Dijkstra [4] famously observed (and cautioned) that a debugged program does not necessarily assure that the program correctly interprets the intentions. This clearly meant that software development and program writing ought to be error prevention oriented. What this means is: Error detection should not be seen as an activity to be initiated after writing a program. Instead, *the program development process ought to embed mechanisms which ensure that errors do not occur, and the intent is correctly encoded.*

A SW is a product. As a product it may be commissioned to be delivered on contract, or it may even be widely distributed for use by users and clients. Therefore, it ought to have a process, organizational management, and quality assurance. In practice, the iterative SW development and testing gradually evolved and found favor with SW engineers. Kent Beck [7] offered following two rules to formulate TDD (Test Driven Development) formalism:

- **Rule 1:** Write new business code only when an automated test has failed.
- **Rule 2:** Eliminate duplications, i.e., use the principles of once and once only.

Closer to the year 2000, TDD found much wider acceptance when it was adapted as a part of eXtreme programming (XP) methodology to generate better quality code. The idea was to generate a code to instill a degree of confidence in its correctness. TDD is often defined as a "risk averse" methodology [8]. The basic philosophy in TDD is to progress step-by-step so that each incremental program passes a designated minimalistic test plan.

The TDD process begins with a clean slate, i.e., a program with a vacuous body and a test that would fail. The next step is to populate the body minimally with some code so that the test passes. As is obvious tests are planned for each anticipated incremental code. Simply stated: TDD stipulates the following repetitive program development steps:

1. Add a test (to fail).
2. Run all tests and see if the new one fails.
3. Write some code to pass the added test.
4. Run tests once before beautifying / optimizing the code.
5. Refactor code (make it clean and understandable or optimize it).
6. Run tests if needed.
7. Repeat steps 1 through 6 till the program is completed.

TDD has two key propositions, e.g., to pass a unit test and subsequently pass an acceptance test when the unit is integrated in the final product. A product's software tests are broadly classified as functional and non-functional in nature. Functional tests check out correctness. The non-functional tests on the other hand check out performance, reliability, and other such indicators.

Cloud-based product testing is done to assess a web or native mobile platform application (such as for IOS or Android) for scalability, reliability, performance, and security. This form of testing is performed on a third-party Cloud computing environment which can host required infrastructure to perform desired tests. Various types of Cloud testing processes enable stress testing software and hardware without the usual constraints of a limited budget, geographical issues, multiple test cases, costs running high per test, etc.

Benefits of Cloud-based product testing include:

1. Eliminating compute environment-related schedule delays for a geographically distributed team, as everyone can be remotely logging in and sharing the infrastructure.
2. Automation of running Cloud-based tests may improve collaboration between diverse teams and minimize activity overlap between the team members.
3. Total cost of ownership of Cloud-based testing automation tools in generally less than the in-house infrastructure, especially, as many tool providers use a "pay-as-you-go" model.

4. Cloud-based testing tools can be accessing anytime from anywhere with Internet access.
5. As a Cloud-based application grows in scale and size, computer resource can scale accordingly.

For example, cross browser testing on a remote Cloud [9] can be highly beneficial as automation tests can be executed in parallel on the Cloud facilities. Thus, testing on the Cloud enhances the speed and efficiency of automated testing and consequent delivery of the product. It is possible to identify performance issues in a Cloud-based product, by scaling the application to a large number of concurrent virtual users.

It is possible to test native web application without requiring real mobile devices (e.g., an iPhone or iPad) by using virtual models of these devices. It is imperative that a Cloud testing platform supports different combinations of browsers, platforms, desktop devices, and mobile simulators.

4.9 Business Driven Development

Note that Agility invokes the use of TDD methodology. In the same vein, some IT professionals reasoned that Agility ought to be brought in the IT backbone of business systems. Behemoths like IBM, Accenture, SAP labs, TCS, and Infosys or banks like Bank of America, HDFC, etc. are in IT services segment or enterprise computing segments. For these organizations, the tests ought to be driven to check the correctness of the process (for its behavior). In other words, one needs to cater to the business process behavioral testing and integration. This assures successful realization of the business objectives in an automated system. This is where BDD [10] comes into picture. BDD is variously expanded as behavior driven design and development as well as business driven design and even business driven testing. However, from our stand point, we shall not make this distinction. BDD, in the spirit of Agility, requires behavioral assurances. Therefore, like TDD, BDD also precludes some definitive steps to be followed. The following steps define the BDD process:

1. Think "behavior," i.e., "tests" are designed to meet the specified behavioral requirements.
2. Ensure collaborative operations among business stakeholder, business analysts, quality assurance team, and the developers.
3. Use a commonly shared vocabulary which easily states behavioral requirements.
4. The exercise is to be driven by identifying the business value that accrues.
5. A SW tool is used to ensure steps 1 through 4 above.

The way BDD is achieved is that the TDD process is extended to ensure that all the stake holders—including non-technical business promoters are on board. Currently, IT community is employing some frameworks that help to connect or bridge the business with the technical infrastructure. One outcome of promotion of BDD has been the extent of automation that has been achieved. BDD is driven by tools like Cucumber or JBehave. These tools enable appropriate communication between the business and technical groups.

The promoters of BDD claim the following benefits:

- Independence from platform lock-ins. This is a huge benefit as this allows within an organization with multiple verticals to choose their IT platforms and operating environments.
- Independence from testing framework design. Basically, with this the test automation processes for specific domains can be created with greater degree of freedom.
- The final benefit one would like to cite is the extent of test documentation that is achieved by automated tools. In fact, higher management levels seem to place greater degree of faith in automated documentation.

BDD tests an application's behavior from its end-user's point of view, whereas TDD is used to test smaller parts of an application's functionality in isolation. A suitable combination of various technologies discussed in this chapter (namely BDD, Agile, TDD, and DevOps) is used when a Cloud-based product is ready to be deployed to clients worldwide. They can access the applications on various servers via load balancers and routers on the Internet, as shown in Fig. 4.9.

Although this is not a book about Cloud security [6], it is pertinent to show an example of Cloud-based security testing. Due to an application's scalability and wide network access offered by Internet to the potential hackers, unique opportunities may arise in the field that may not be foreseen in a controlled deployment environment. We will revisit blockchain and its security implications. There are several published examples of hackers stealing significant amounts of crypto-currencies from the exchanges. However, it is important to state that all known breaches are not in the blockchain itself, but in the applications built on top of it as is evident from the following incident which was widely reported.

In a recent incident [11], an attacker somehow gained control of more than half of the Ethereum Classic network to rewrite its transaction history. That enabled the hacker to spend same cryptocurrency more than once, known as "double-spend." This was accomplished by the hacker controlling more than 50% of Ethereum network's computing capacity. This provided an opportunity to defraud other users

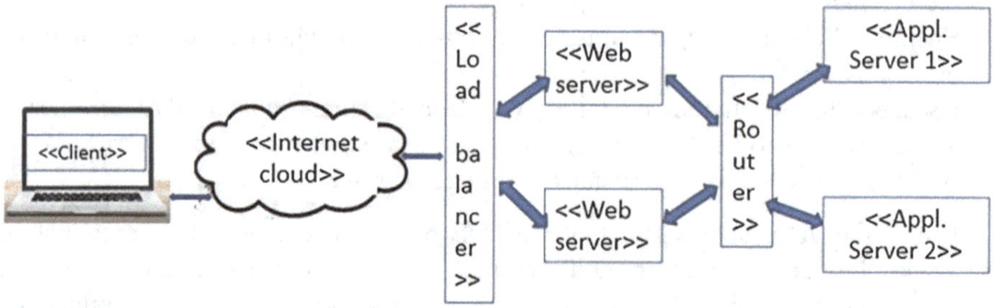

Fig. 4.9 Cloud-based deployment of a web-enabled application [4]

by paying them in cryptocurrency, before creating an alternate version of the blockchain in which this payment never happened. Eventually, the alternate version was voted by the hacker to become the authoritative blockchain so the same currency could be spent again. Its system administrator Coinbase was able to pause interactions when it noticed the problem with the compromised blockchain.

Thus, a solution that works in theory may breakdown in practice if any of the key assumptions in a project are violated. This problem can be addressed by setting up automated monitors to flag any unusual or suspicious activities in a Cloud-based application.

4.10 Product Roadmap Management

Program managers and senior executives typically maintain two sets of roadmaps: a short one for the current product under development, which shows a detailed schedule for design, verification, validation, and product launch at hand. The second one usually is a multi-generational roadmap for next 2–3 years is maintained to show how the major features or big-ticket items are lined up over time. This is important to keep the team's morale up as well as to keep the management informed for future resource planning.

When a team is executing the current project, and some of them roll off at the end of their development cycle, the developers need to know what is in the pipeline for next generation. In fact, there are cases in the Silicon Valley when the developers of major multi-year projects start looking for opportunities elsewhere. Similarly, if the next program is more complex and requires new skills, such as virtualization or support on a new device platform, then resources with appropriate skillset need to be hired in advance. An example is a Progressive Web App (PWA) [6, 8], which can run in a range of browsers including on mobile devices. However, after an initial release, marketing found that users prefer a native App running on IOS or Android devices. This could be due to a need to work in offline mode, which PWA could not support. In addition, users often desire better integration with a device's native operating system. In either case, project managers will need to plan in advance and hire developers with the native device development skillset. Furthermore, they need to carefully manage the future releases with equivalent features on both sets of applications. Thus, we recommend that not one but 2–3 roadmaps may need to be maintained.

In general, a roadmap is a living document that changes based on the users' perceptions and feedback, market conditions, and competitive product offerings. Not being able to react quickly and adapt to changing requirements has driven many good products to demise even when it was backed by an otherwise successful company like Google [12]. The case in point is about Google+, a social network created in response to Facebook's growing popularity. While Google's massive reach helped it to build a sizeable user base for its new social networking product, its existing users were signing up but not using Google+ (or G-Plus). The company even merged its other product lines such as Google Photos and YouTube to auto-share content in support of G-Plus product. When a bug was discovered in G-Plus, which exposed personal data of nearly half-a-million users, prompting Google to kill the fledgling product line [13]. In this case, social networking customers and marketplace voted for Facebook with their usage and advertising wallets. This offers an important lesson in building products that appeal to the users. *Success in an existing domain (search engine in case of Google) does not automatically translate to success in the other areas.*

4.11 Engineering Change Order (ECO)

Regardless of a Cloud-based development methodology, a product's requirements are likely to change over time. Changes may come from customers, competition or internally driven.

- **Customer-driven changes:** Sometimes end users are not able to articulate their needs upfront. This is especially true for the new products, such as iPhone development. Apple knew that users want to integrate a phone, camera, and other web-enabled applications but exact engineering requirements for iPhone's GUI (Graphical User Interface), communication protocols, product's form factor, and charging methods were not clear. Software typically goes through a sequence of alpha, beta, and production releases. These evolved by building early prototypes, showing it to potential users and partners, launching products, learning from these to build the next generation of iPhones in an iterative manner. That is how later iPhone models grew in the screen size, dropped a physical home button and added face recognition etc.
- **Competition:** Sometimes a product team sets out to build with some initial requirements. However, unknown to them is the work being done by the competition which may be ahead to launch a better product. This causes a disruption in the product plans as customers are unlikely to buy an inferior product with fewer or less efficient features. An example is how Apple's iPhone won unprecedented mainstream success over Blackberry's RIM [14]. The latter had evolved from a pager, as a way for people to access their email on the road. It had a keyboard, while the former evolved from an iPod with a multi-touch screen with no physical keyboard. Apple's full-screen interface won over customers with its simple and easy to use product.
- **Technology driven:** As they say in the technology industry that change is the only constant. Technology is sometimes driven by its own needs without explicit user input. An example is how Google's Android operating system overtook Nokia's Symbian [15]. Android was designed for touchscreen phones or devices, while Symbian was designed mainly for keypad phones. Symbian was released earlier than Android and had a larger market share. Despite the first mover advantage, Android won because it was based on Linux and could do multi-tasking.

Android is full of features, and mobile phone companies compete to manufacture the best Android smartphones. In contrast, Symbian failed to keep up with the technological changes and lost its place in the market. Sometimes, a product design team may come up with new ideas, which the sales and marketing teams then test in the field to see if users will buy such a product.

The expected output from the requirements change process [4] should help to evolve specifications that could lead to an effective and useful design. Also, it should be cleanly documented and presented in a manner that facilitates defining the software architecture. It helps to iterate and update requirements. The iteration is necessary for the following reasons:

1. Market observation as it is likely to evolve.
2. New use scenarios observed or trends in usage pattern.
3. Removing redundancy from overlapping set of "Requirements Statements."
4. Eliminating ambiguities or contradictions in "Requirements Statements."

Independent of the source of changes in the requirements, a product team needs a clear process to accept changes in the requirements, then judge if the changes sought can be incrementally integrated into the current product or must wait until the next release. A typical way to decide this is by ROI (Return on Investment) approach. It simply compares the cost of making a change in terms of time and effort required, vs. the customer value of such a change. ROI for an ECO change is based on the stage of the project. Some changes are non-negotiable such as required for security. An example is a Cloud Application to be used by doctors in a hospital. The IT department for that hospital may be using a Single Sign on (SSO) process, so separate username and passwords are not required in every application. This is to ensure that when a new doctor joins, or one quits, then their account can be added or deleted from a single database. This requirement can be a sales blocker, i.e., the hospital IT will not authorize deployment of the App unless it can support SSO using their specific vendor. Then the design team has no option but to incrementally work it into the current product.

The point on change management needs elaboration [4]. While a change in requirements ought to be permissible, it must adhere to a change management protocol. This is because the document itself results from considerable churn. Clearly, the protocol ought to cover the following:

1. A full justification of why the change is needed.
2. Form of the change required.
3. Identification of the requirement (may be requirement id or no.)
4. Priority assignment.

When a change request is received, the persons who own the requirement and its scope need to be consulted. The change can be incorporated only through consensus. An example is a change control process that involves representatives from Marketing, Engineering, Quality Control, and Sales departments to assess the ROI (Return on Investment) for the requested change. An outcome of this process can be no change, an alternation in the currently ongoing project, or delay the requirement till a follow-on product release.

An example is a requirement to add a new feature to the Cloud Application, such as adding a new color icon, which is deemed nice to have but can clearly wait till the next product revision. However, another requirement to save the report in a certain format for customers to use can be a sales blocker. In this case, former is delayed, whereas the latter is accepted in the current product's release cycle. Similarly, requirements that may require major architectural changes need to be planned in advance. Typically, a separate PoC can be done to assess the technology readiness and feasibility of a new feature, before it is worked into a product's plans.

4.12 Partner Management

Gone are the days of a dedicated small team working in a garage to bring out revolutionary products [16]. Most software products these days require teams spread across multiple time zones. This is to harness the talent located across the multiple geographies and also a desire to be closer to the global markets being served. While web enables such development efforts making the collaboration easier using online tools, the project management tasks become challenging and different. Cloud-based software products are also expected to work on a variety of devices (laptops, smartphones, etc.) using different operating systems (Windows, IOS, Android, etc.) and browsers (Chrome, Edge, Mozilla, Safari, etc.). Such interoperability requirements increase the cost of development and testing. Furthermore, software products deployed in specialized environments such as hospitals must follow the relevant regulatory processes and must adhere to protocols such as HIPPA. These require extensive documentation for approvals such as Software as a Medical Device (SaMD) by FDA [17].

A software product exists in an ecosystem already populated by other devices, so standards and protocols compliances are needed. For illustration purposes, if we look further into SaMD products, not only the FDA approvals are needed but hospitals customers may want these products to comply with additional standards such as DICOM [18] and HL7 [19]. DICOM enable medical image acquisition devices to transmit, store, retrieve, print, process, and display data from different device manufacturers. Similarly, HL7 is an ANSI-accredited standard for exchange, integration, sharing, and retrieval of electronic health information.

To accomplish business success for a new software product, developing it may be easier than getting it accepted in the existing ecosystem. Success for it needs to be planned upfront at the requirements gathering stage, with inputs from domain IT managers, users, and other stakeholders, which may include other product vendors in the users' workflows. This has given birth to a new industry term, Frenemy, which is an oxymoron and a portmanteau of "friend" and "enemy" in the business parlance [20]. An example of that is how Netflix may view Amazon, as the former offers a SaaS platform for delivering movies and entertainment content to the customers. It uses latter's IaaS platform to offer digital content, but Amazon also has its own Prime platform that offers movies and entertainment content of the customers. Hence, both companies collaborate and at the same time compete for customers' wallets and time. Netflix traditionally was a DVD renting company, which migrated to Amazon Cloud in 2008, after experiencing a major database corruption and for 3 days could not ship DVDs to its customers [21]. To avoid a single point of failure, such as with relational database in its own datacenter, it adopted Amazon Web Services (AWS), which offered a more reliable, horizontally scalable Cloud and distributed resources.

After 7 years, it completed the migration and closed its own datacenters. This enabled Netflix to service 8× the customer base, with three orders of magnitude growth in the overall content viewing, as shown in Fig. 4.10. However, Netflix's migration and Cloud application were tightly tied to AWS, so later on when Amazon started to compete by offering its own Prime services. a. migration to another Cloud service provider was ruled out. *A lesson here is: plan upfront for a multi-Cloud strategy and avoid complete dependence on a single service provider.* This may require an architectural abstraction layer between the Cloud application and APIs provided by the service provider.

These days partner and relationship management terms are used interchangeably, and even together. There are applications available for Partner Relations Management (PRM) [22]. This is more akin to stakeholder management, and their goal is to streamline, enhance, amplify, and strengthen business processes with indirect partners to deliver mutual growth. PRM products [23] provide visibility into and influence over the complexity of a product with multiple channels, such as a Cloud App which needs to work with multiple devices in a hospital, aggregate data to draw inferences and then convey these to the medical care team, some members

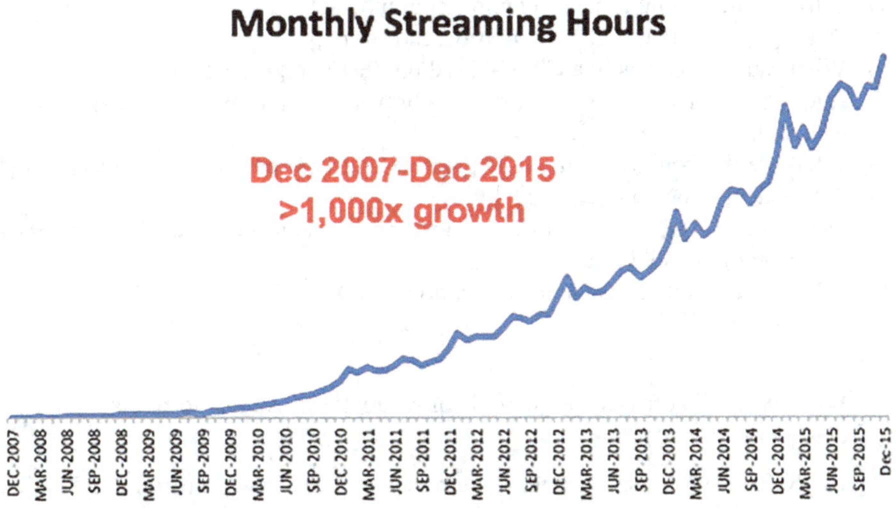

Fig. 4.10 Subscriber growth for Netflix viewers supported by the Cloud [21]

of which may be outside that hospital's firewall boundaries. An App like this has many user profiles, different requirements, and needs to comply with many regulations. We will study the development cycle for one such Cloud application later in the book.

4.13 Summary

The software development process is complex, and it is not always possible to know all the current and future requirements upfront. Thus, its architecture and consequent development need to evolve over time for business adoption and market success. Agile and iterative processes are better suited than a Waterfall model to develop software products. Furthermore, the development team may be split across multiple geographies. In such a case, the communications and expectations need to be managed dynamically. At the same time, a Cloud application product should not be tied to a single service provider as we saw in the example of Netflix, where its service provider started to compete with its own customer. A multi-Cloud strategy is desirable for any new Cloud software project. It is important to test a Cloud-based product with a Cloud-based testing platform for scalability, reliability, and performance on a variety of browsers and mobile devices.

Points to Ponder

1. What type of software projects, if any, are better suited to adopt Waterfall development model?
2. What are the pitfalls, if any, for the Agile software development process?
3. What is the value of doing a PoC?
4. Why it is beneficial to have a multi-geography software development team?
5. Why is it important to have a multi-Cloud strategy?
6. What causes changes in a Cloud-based product's requirements?
7. If a web-based application can run in a browser on a mobile device, what's the need to have native applications?
8. What are the benefits of a CICD pipeline and how can one find errors found during deployment be resolved?
9. DevOps and Agile are two different and independent processes. How can they help a team to collaborate?
10. Which one is better to test an application: TDD or BDD?

Answers

1. What type of software projects, if any, are better suited to adopt Waterfall development model?

   (a) Software projects where requirements are known upfront and using established technologies may be suitable for Waterfall development model.
   (b) A new project where prior experience can be replicated.

2. What are the pitfalls, if any, for the Agile software development process?

   (a) User Waterfall model if the conditions of answers to Q1 above apply.
   (b) If a small software project's features and work cannot be divided into smaller chunks, then it is not suitable to use Agile methodology.
   (c) Agile also required constant interaction with the users at the end of each iteration to get their feedback. So, if the user feedback is not available, then project may get driven in a potentially wrong direction.

3. What is the value of doing a PoC?

   (a) A proof of concept shows the feasibility of a new or complex idea. This demonstrates that the underlying technology works so that it can be used in the main project.
   (b) Assumptions need to be carefully validated and documented because if they change between a PoC and the main project, then the results of PoC will not hold in the final product. In some sense, a PoC assures all stake holders about the correctness of assumptions made.

4. Why it is beneficial to have a multi-geography software development team?

   (a) Besides the potential to lower the cost by locating a development team in a more affordable setting. This will enable a project manager to harness global talent. Basically, the strategy is to take the SW development work where the talent is available with cost arbitrage.
   (b) A team located closer to the future customer base can understand their needs better and react faster to the bug fixes or changes needed.
   (c) Different time zones effectively extend work hours during a day.

5. Why is it important to have a multi-Cloud strategy?

   (a) A situation where the Cloud service supplier can drive tight integration of its customer's software product to its platform, the in future can change the pricing, or lower availability, which is not desirable.
   (b) Another limitation of tight integration with a Cloud service supplier's platform is if the latter decides to compete with its customer's products, then customer has no options.
   (c) Thus, it is desirable for a Cloud-based software product to support multi-Cloud strategy, to optimize for cost and moving to a friendly ecosystem.

6. What causes changes in a Cloud-based product's requirements?

   (a) Besides the usual drivers of change for product requirements changes, namely competition and user needs, Cloud-based applications may need to scale as and when the number of users grows or shrinks.
   (b) A Cloud-based application's architecture needs to be scalable without any discernable impact on its performance. Take the example of Facebook or Youtube, as its users upload more and longer videos online for other subscribers to watch.

7. If a web-based application can run in a browser on a mobile device, what's the need to have a similar functionality be available as a native device application?

   (a) Generally, every mobile device has a browser in which a Cloud-based application can run. However, due to security requirements, the mobile devices do not allow the browser to auto-save files on the native device.
   (b) If the users of a Cloud-based application wish to access its data in offline mode without Internet, e.g., during an airborne flight, then a native application is needed.
   (c) An example is Netflix application to view movies, which many users want to access on their mobile phones and iPad devices during long airplane trips. This is not possible on Netflix's web browser-based application, so a native mobile device version is needed. A native application is often downloadable on the mobile device provider's web site.

8. What are the benefits of a CICD pipeline and how can one find errors found during deployment be resolved?

   (a) A CICD pipeline automates the build, test, and deploy phases of the release process every time, there is a code change for a new release. There is a potential for an error to escape if the test coverage is not adequate. In this case, the deployed version can be rolled back to a previous release, while the bugs are fixed and code is released. One needs to ensure that the regression test cases continue to grow for known bugs in all the previous versions.

9. DevOps and Agile are two different and independent processes. How can they help a team to collaborate?

   (a) For Cloud-based products development, a development team may follow Agile process in one location, while an operations team in another location can use DevOps to make regular releases to customers spread across the world. Thus, the two teams can work collaboratively by providing feedback to each other.
   (b) Such an arrangement promotes some cooperative competition between these two teams to meet each other's expectations and schedules.

10. Which one is better to test an application: TDD or BDD?

    (a) TDD and BDD are not mutually exclusive. BDD tests an application's behavior from its end-user's point of view, whereas TDD is used to test smaller parts of an application's functionality in isolation. We would like to remark here that the orchestrated encore of the behavior is determined by the encoded functionality.
    (b) Some Agile teams may use TDD only to ensure functionally correct programs. Whereas BDD ensures that the application works as intended. So, both are needed.
    (c) One may consider TDD as a separate unit level testing to check that each component of an application works properly. In that case, BDD can be considered as an integration level test that shows that the whole application works as intended.

References

1. https://bitcoin.org/bitcoin.pdf
2. https://www.tutorialspoint.com/sdlc/index.htm
3. SAP 01: manuals for program leaders (2005).
4. Bhatt, P. C. P. (2021). *Software design, architecture and engineering: Concepts and practice*. Prentice Hall India, Easter Economy Edition. ISBN-978-81-951611-4-0.
5. https://Agilemanifesto.org/history.html
6. Sehgal, N. K., Bhatt, P. C. P., & Acken, J. M. (2023). *Cloud computing with security and scalability*. Springer. https://link.springer.com/book/10.1007/978-3-031-07242-0
7. Beck, K. (2002). *Test driver development: By example*. Addison-Wesley Professional. ISBN 978-0321146533.
8. Budhabhatti, M. (2008). TDD and continuous integration for mobile apps. *The Architecture Journal, 14*.
9. https://www.lambdatest.com/selenium-automation
10. Bhatt, P. C. P., & Mueller, W. (2000). *Morphic nets: Model based Design diagrams*. UKPEW, Univ. of Durham.
11. https://www.technologyreview.com/2019/01/08/137908/hackers-just-stole-1-million-from-the-ethereum-classic-blockchain-in-a-rare-51/
12. https://killedbygoogle.com/
13. https://mashable.com/article/google-plus-shutdown
14. https://www.imore.com/how-apple-killed-blackberry
15. https://askanydifference.com/difference-between-android-and-symbian/
16. https://www.hpmuseum.org/garage/garage.htm
17. https://www.fda.gov/medical-devices/digital-health-center-excellence/software-medical-device-SaMD
18. https://www.dicomstandard.org/
19. https://www.hl7.org/
20. https://en.wikipedia.org/wiki/Frenemy
21. https://about.netflix.com/en/news/completing-the-netflix-Cloud-migration
22. https://www.gartner.com/reviews/market/partner-relationship-management-applications
23. https://www.cs.utexas.edu/users/EWD/transcriptions/EWD02xx/EWD215.html

Chapter 5
IoT-Based Application Development Process

5.1 Introduction

This chapter focuses on IoT product development and management processes as distinct from the software development and management process discussed in Chap. 4. The discussion begins with the requirements gathering and continues through testing and deployment of an IoT product. Specific details are given to the hardware and embedded software development in an IoT product. Special emphasis is given to the unique challenges in testing IoT products during development.

5.2 IoT System and Subsystem Requirements Gathering

The discussion on requirements gathering and engineering from Sect. 4.4 is applicable across all disciplines and modes of product development. As internet-connected hardware devices, IoT systems and subsystems introduce additional considerations throughout the requirements development process [1]. Those subsystems may include electronics, embedded software, communications (i.e., what protocols are used and in what fashion) and mechanical subsystems. Given the high cost of iterating on hardware (electronics and mechanical) designs, it is important to understand the challenges in capturing meaningful requirements throughout the project life cycle. Some important aspects to consider are subsystem teams' organization, domain knowledge over many fields, and requirements readiness versus subsystems. This entails generating specifications across multiple interconnected domains.

Based on the architecture chosen for an IoT system, different engineering teams working on the project can have different requirements engineering responsibilities. For some products, it may be necessary to tightly couple hardware and software

P. C. P. Bhatt et al., *Project Management in Cloud and IoT Applications*,
https://doi.org/10.1007/978-3-031-99427-2_5

requirements. This category includes products for which the software implementation has a high reliance on hardware processing and sensors and vice versa. For example, a self-driving car relies on a suite of specific hardware sensors for which specific software processing is necessary. While such products can (and should) be designed to allow for some amount of team independence, there will be scenarios where updated requirements must be simultaneously implemented across many teams. For other products, the hardware and software requirements can be mostly decoupled from each other. This is the case when a strong Interface Control Document (ICD) is developed for the product that defines the required method of communication and interaction between various subsystems. An example is the NovaGuide system [2]. In this system, the NovaGuide View Cloud application receives all necessary data for display and analysis through an encrypted Internet connection from the NovaGuide ultrasound machine. The NovaGuide View Cloud application team is free to iterate on their requirements engineering for data analysis and display without the need to modify the requirements of the NovaGuide ultrasound machine. Further details of the application based on the NovaGuide system will be discussed in Chap. 7.

To ensure a successful project, the requirements engineering team should have a broad understanding of the technical fields and domains encompassing the project. This allows them to assess and translate customer needs into requirements that can be met by the engineering teams and verified by the test teams. A Cloud application project may only involve software and networking development; however, an IoT project typically involves software, hardware, electronics, and networking development. The responsibilities and accountability of the requirements engineering team for an IoT project is thus greater than the responsibilities of a project of smaller scope.

As noted for any project [3], requirements volatility is a challenge and can occur for many reasons. Three common reasons cited by the Standish group are:

- Lack of user input: 13% of all projects.
- Incomplete requirements and specifications: 12% of all projects.
- Changing requirements and specifications: 12% of all projects.

The above issues are compounded in IoT projects where both the hardware and software developments are enmeshed from start to finish. For hardware and electronics, translating requirements into a design and manufacturing a circuit board can take many weeks or even months. This slower development cycle is particularly impacted by incomplete or changing requirements. Based on the requirements changes, new hardware may need to be developed which could jeopardize the project release schedule. The following examples are typical of hardware requirements and how an electronics design may be impacted when the requirements change:

- The product shall be powered from an external AC/DC power supply.
  - Such a requirement puts the burden of the AC/DC power converter and safety circuits onto the AC/DC power supply chosen. Further, it dictates the product power port form factor. If it is later decided that the product should plug directly into the wall for power (like a personal computer), a power electronics

engineer may be needed to design or select the correct internal AC/DC converter. The enclosure form factor would also change to include the space and connector for the AC/DC converter.

- The product shall include Wi-Fi (802.11) connectivity.
  - A Wi-Fi connectivity requirement would dictate the internal wireless modules used in the device as well as regulatory considerations for use of electromagnetic spectrum. A change to wireless communication requirements might not only drive a component change but also entail an additional wireless testing at the verification phase.
- The product shall include a mechanical failsafe button to stop operation.
  - Mechanical failsafe buttons are typically large, obvious, and easy to access. As such, they require careful integration into a mechanical design for robustness and usability. They may also require a standalone circuit to ensure operation when the primary product circuit is failing. If, late in a project, it is decided a mechanical failsafe is not necessary, the cost and resources for the mechanical and electrical design work done would be of little value.
- The product shall display operational status with a collection of LED indicators.
  - This could be satisfied with several simple circuits driven by general-purpose input/output (GPIO) pins available on many microcontrollers and minimal firmware support. If the display content becomes more complex and the design is changed to require an LCD screen, the electronics may be greatly impacted. An LCD screen could require a fast serial communication interface such as SPI and associated firmware drivers. If the initial microcontroller chosen did not have sufficient resources, the entire circuit may need re-design.

All the challenges faced in IoT system design described above can be effectively mitigated. This is one of the responsibilities of the project manager. At the onset of a project, the manager should ensure that all appropriate engineering teams for an IoT project are fully resourced. These engineering teams should discuss communication strategies, both in terms of day-to-day communication within the product team and for technical communication between product subsystems (ICDs). This should be done upfront and at the time of the requirements engineering process. To address the challenge of capturing requirements and specifications across multiple interconnected domains, the project managers must appropriately staff the requirements and systems engineering team to include all relevant disciplines in the project. This may introduce a challenge when it comes to budgeting for the project since a larger systems engineering team may be needed. It is important for the key corporate stakeholders in the project to understand the long-term cost-benefit of poor requirements engineering. Similarly, the cost of requirements volatility for an IoT must be appreciated. Project managers should drive to identify key aspects of hardware and network requirements (e.g., data bandwidth) early in the project over more malleable software requirements. Doing so can avoid delays in the project caused by long lead hardware development cycles.

5.3 Understanding Sensors

Sensors play a key role in IoT products. They translate information from the real world into digital information. The types of sensors available are wide ranging, but all types have a common purpose of allowing electronics and software to interact with the environment. Table 5.1 lists various types of IoT sensors used in consumer electronics, civil, industrial, and healthcare applications. This is not an exhaustive list of sensors or applications, but it should be useful for understanding the typical sensors an IoT product design may need to support.

Regardless of what specific sensors are used in an IoT product, they must be integrated into the system electronics by a hardware development team. Power and analog front-end electronics are skills often needed while working with sensors. Sensors can be passive or active. Passive sensors do not require circuitry to apply power to the sensor to read a valid measurement. For example, a ball-in-tube mechanical crash sensor in a car uses a physical trigger powered by a vigorous impact and does not necessarily need electronics for sensing a crash. Active sensors require a power source before a valid measurement can be taken. A camera or CMOS sensor needs a power supply and must be turned on before light levels can be measured. For both passive and active sensors, a common method of taking a measurement is with an Analog-to-Digital Converter (ADC) integrated circuit chip. These chips digitize the voltage or current values provided by a sensor and are necessary to allow microprocessors or CPUs to act on the data. There are thousands of

Table 5.1 Types of IoT sensors

Sensor	Typical devices/uses
Temperature	Medical and personal body temperature measurement, manufacturing equipment monitoring, home thermostats
Humidity	Personal and professional weather stations, heating and air conditioning systems, home thermostats
Image	Facial and pattern recognition, object ranging and detection, night vision, computer vision
Infrared (IR)	Object detection, heat sensing, remote control, blood pressure, and blood flow monitoring
Proximity	Cell phones, smart watches, self-driving cars, manufacturing inventory control, parking meters, Bluetooth, and NFC communication devices
Touch	Cell phones, smart watches, human interface devices (HID)
Smoke detection	Personal and professional fire detectors, industrial process monitoring
Water quality	Water distribution systems, pH monitoring, turbidity monitoring
Ultrasound	Flow rate monitoring of fluids, noninvasive imaging, blood flow monitoring
Pressure	Leak testing, gas or liquid monitoring in vehicles or aircraft, traffic light, and parking management
Position	Cell phones, GPS devices, velocity, and acceleration tracking
Soil moisture	Crop and irrigation management, greenhouse productivity

ADCs available to use with different parameters such as resolution, speed, and input amplitude. Signal conditioning electronics is also often needed to ensure that the sensor signal connects to the ADC in a way to maximize signal-to-noise ratio. *When working with sensors, the electrical engineering team should have a robust understanding of more than just digital electronics.*

Once the sensor has been integrated into the electronics subsystem, the data from that sensor can be processed digitally. By transmitting the data collected with an IoT product to a Cloud application, all the benefits of the Cloud amplify the ability of the product while simultaneously reducing the embedded processing capability needed by the IoT electronics. Embedded device electronics by itself are severely constrained in memory and processing power compared to the Cloud. As such, IoT products that are not connected to the Cloud can only provide limited features. A home smart light may be able to automatically turn on and off based on time of day or sound detection when not connected to the Cloud. However, when connected to the Cloud, the smart light could be turned on and off from anywhere in the world or could predict desired light levels and timing based on machine learning. The convenience of this may be tempered with issues such as Cloud latency, but products can always keep critical functionality solved on the device. When considering networks of IoT products, the scalability of the Cloud introduces another key benefit. Cloud connected IoT nodes in a network can have a single policy applied to any or all relevant nodes across a wide swath. Alternatively, a distributed management policy may be more appropriate. Such could be the case for a network of electrical power grid demand monitors. As more IoT nodes are added to the network, the upkeep and application of policy rules remain minimal with the Cloud.

5.4 Prototyping Hardware

IoT devices and products necessarily include hardware and electronics. Project managers must be familiar with methods of prototyping hardware both to ensure project success and to fully estimate the effort involved in developing completed electronics boards. This is the hardware analog to the Proof of Concept (PoC) discussed in Chap. 4 for a Cloud Application. Some aspects of prototyping hardware circuits and Field-Programmable Gate Array (FPGA) circuits are presented.

5.4.1 Hardware Circuits

5.4.1.1 Hardware Circuits: Simulation

Simulating analog and digital circuits serves as an important first step when designing new hardware circuits. Software simulation allows the engineer to easily compare a circuit's performance with its intended performance.

It is important to note that simulations are only as accurate as the models from which they are built and that no model perfectly resembles real-world components. Simulation results should not be used to judge the ultimate performance of a circuit. Rather, prototypes should be constructed to evaluate a circuit after simulation.

5.4.1.2 Hardware Circuits: Prototyping

There are several methods to constructing prototype circuits. The appropriate prototype method must be selected based on the type of circuit being designed. If an inappropriate prototype method is used, the circuit may exhibit faults that are not due to circuit design but caused by the poor construction of the prototype.

For digital circuits, an off-the-shelf solderless breadboard may be used. When using solderless breadboards, interconnects must be made with wire of the appropriate gauge. Interconnect lengths must be kept to a minimum and should be color-coded to indicate functionality. Interconnects and components should be placed neatly, avoiding crossing or bunching wires. This allows the engineer to more easily compare the breadboard prototype to a schematic and increases testability. Solderless breadboards may also be used for analog circuits involving such low frequency signals (<1 MHz) where signal fidelity is not crucial.

For analog circuits involving high frequency signals (≥1 MHz), high fidelity low frequency applications like audio, or for digital circuits involving very high frequency signals (>100 MHz), prototypes should be made with a surface mount or through hole breadboard that has a ground plane. Components should be connected with an effort made to keep lead lengths as short as possible. This method should only be used if the number of interconnects and components is small.

When prototyping an analog, very high frequency digital circuit or any circuit that is too large in component and interconnect count to hand assemble in a reasonable amount of time, the engineer should create a printed circuit board (PCB) for test. Good design and layout rules should be followed when creating PCBs, even if the boards are only for prototyping.

For complicated integrated circuits, an evaluation board must be purchased and used if possible. This saves the engineer time in designing and/or constructing a prototype to evaluate the integrated circuit.

It is important to design hardware circuits in a way that they can be tested, either manually or in an automated fashion. The following techniques are used when prototyping circuits:

- Create test points throughout the circuit. Test points are single pins or large vias that can be used to probe desired locations within a circuit.
- A circuit is typically divided into functional blocks which are interconnected. Measurement and functional access points should be built into the circuit in between the functional blocks. Usually, the functional access points include a physical interface that allows a tester to inject a test signal into the circuit's functional block under test.

- Commonly, in more complicated integrated circuits, manufacturers include integrated test functionality. These functions may require custom hardware or firmware to enable this test functionality. Always expose this test functionality on the PCB when prototyping.
- Built-in Self-Test (BIST). BIST is a method for electronics to test itself. BIST can be at power on (POST) or while the device is powered but in a standby state. As an example, a voice activated smart home speaker system may create tones when powered on. The system can internally use those tones to test that its microphone is working for voice recording. BIST is often designed where scripting tools can be used by engineers to create various test scenarios and success conditions with the embedded hardware available.

While it is not necessary to maintain all test points or functionality in final revisions of production quality circuits, thought should be given to keeping the appropriate testability to ensure the circuit is properly manufactured.

5.4.2 Field-Programmable Gate Array (FPGA)

FPGAs are commonplace in complex digital circuits. They provide low latency parallel processing which is often more difficult to perform in the sequential processors available on microcontrollers. As such, applications in which a large amount of data needs to be processed in real-time control environments may use one or multiple FPGAs. They are also re-programmable in the field which makes them valuable for product maintenance but also the prime candidate for prototyping Application-Specific Integrated Circuits (ASICs). ASICs are more power efficient than FPGAs and best suited for particularly intensive processing which is well defined and is not expected to require updates in the field.

FPGAs have JTAG capability, which gives them inherent design for test functionality if the hardware design has provided an interface to JTAG. In addition to JTAG:

- Simple test logic can be written to aid initial bring-up and can be left within the system if it does not significantly impact FPGA build times and timing.
- Counters can be placed along the data path to check that various points along the data path behave as they should.
- Inputs to the FPGA should be checked for validity and the FPGA should indicate errors in inputs via interrupts or other means.

5.4.2.1 FPGA: Behavioral Simulation

FPGA code functionality is mostly verified through behavioral simulation. Behavioral simulation provides complete visibility at a level that is easiest to understand, but due to the highly parallel nature of digital circuits, simulation of complex circuits can still be very time-consuming. Based on the complexity of the functionality, the simulation can vary from a quick sanity check that is manually verified to a thorough, self-checking testbench with appropriate supporting test stimulus. Behavioral simulation will generally not expose timing-related behavior.

5.4.2.2 FPGA: Post Synthesis or Post Place and Route Simulation

Post synthesis or post place and route simulation is annotated with timing information from the synthesis or place and route timing analysis. These simulations are extremely time-consuming, sometimes quite difficult to interpret. They are generally not necessary if good design practices are followed. However, if there are issues that appear to be timing related, it is recommended that this simulation methodology be considered.

5.4.2.3 FPGA: Unit/Integration Tests

Integration tests are tests on the actual FPGA controlled by software. Since it is time-consuming to simulate many permutations even behaviorally, integration tests are much better suited to thorough testing of many combinations of inputs in real-time. However, due to the lack of visibility in the actual FPGA, problems found in an integration test should be simplified so that they can be reproduced in behavioral simulation. If this is not possible, debug logic can be added to the FPGA to debug in situ.

FPGA engineering is supported by a growing toolset to aid in design capture and test. Vivado Design Suite [4] provided by AMD (AMD acquired the FPGA company Xilinx) and Quartus Prime Design Software [5] provided by Intel (Intel acquired the FPGA company Altera, which since has been spun off) are the development suites for two large suppliers of FPGAs. Within these ecosystems, many intellectual property (IP) cores are available which include sophisticated test benches that are ready-to-use. Best design practices are available for FPGA development and test [6, 7]. It is worthwhile for design teams to review, agree upon and enforce a particular set of FPGA design practices to ensure the highest quality output.

5.5 Embedded Software for IoT Products

In addition to good familiarity with hardware development, the project manager ought to understand the nuances of embedded software development for IoT products. Some of the basics of software development, if neglected, can have a large impact on project timelines and success of the project—given the challenges faced in embedded software development. While many microprocessor and sensor suppliers provide Board Support Packages (BSPs) or Application Programming Interfaces

(APIs), the quality of this supplied code is varied and often leaves much to be desired. It is important for the project manager to plan and allocate extra time to address the challenges when integrating or rewriting these software packages.

Processing power, memory, and peripheral functionality are all constrained in embedded systems compared to Cloud-based systems. For this reason, the fundamental approach to understanding requirements, designing and architecting software prior to coding without these guiding elements in place becomes critical. In embedded systems, painstaking care must be taken when coding to ensure proper operation. Code must be in the correct processor memory locations. Embedded communication peripherals, such as Inter-Integrated Circuit (I2C) and Serial Peripheral Interface (SPI) must be configured with the correct setting and routed to the correct pins on a microcontroller. Functions need to be optimized for execution speed. Concurrent, multi-tasking, and multi-threaded processing require careful coordination to evaluate the perils of pre-emption [8]. When coding starts with a poor or incomplete design, a common result is many rounds of refactoring throughout the project. The challenges listed above may entail a long and complicated refactoring.

High-quality processes should be maintained through the entire embedded software development cycle, not just at the initial upfront design stage. Even with the best architecture and design, if best practices are not followed during coding and test, then the project will likely exceed schedule and budget [9]. Frequent demonstrations and checks into the quality of the embedded team's output can help a project manager assess the health of the project. In addition to Agility, IoT projects follow a paradigm where DevOps are also required for project success.

One aspect an embedded system developer frequently encounters when designing for specific processors, sensors and peripherals is vendor supplied code. It is prudent to have a healthy skepticism of such code. Quality, portability and functionality should all be assessed as soon as a particular set of vendor code is considered for a project.

For best results, the same quality metrics being applied to internally developed code should be considered for vendor supplied code. While some vendors may supply high-quality code, others may not. Even with a single vendor, some code may be high quality while other code is not. In all cases, it is not possible for the IoT project team to know the specifics of the vendor's software development process. As such, the project manager should budget time to assess the quality of vendor supplied code. Simple methods such as code reviews and static analysis can go a long way to improve confidence in the code. For critical components in the design, the project plan should also have contingency plans to modify hardware selection if the vendor supplied code is insufficient.

Another aspect to consider with vendor supplied code is portability. Oftentimes, the code is written and optimized for use with other chipsets from the same vendor. For example, a sensor from ST Microelectronics is likely to provide an API that is designed to work with ST Microelectronics processors. In many cases, IoT projects may require the use of a particular vendor for the microprocessor but different vendors for the sensors or peripherals. These scenarios can introduce portability challenges. If an API is provided for a sensor that assumes a minimum addressable unit of 8-bits, then that API is unlikely to work for a processor that has a minimum addressable unit of 16-bits. If the API assumes a little-endian processor, then that API is similarly unlikely to work on a big-endian processor. These challenges can be overcome but should be planned for as part of the IoT project.

Lastly, vendor supplied code may not always include all the functionality an IoT project needs. While a minimum set of features may be exposed with an API, there could be a considerable amount of work the project team needs to do to take full advantage of the peripheral device. The project manager should not assume that a supplied API includes all the functions necessary for the project. Instead, a careful review of the API and documentation can mitigate any misunderstandings of the feature set provided.

5.6 Design for Low Power Operations

There are several challenges associated with developing a successful IoT product. One challenge is to consider low power design at the beginning of a project, as it is difficult to retrofit a design that was not planned for low power operation. Depending on the application, the device may need to be designed for rugged environments, lack of easy access for maintenance, or limited connectivity to wireless networks. One such challenge is the ability to operate without a large source of power. Some IoT products may have a small battery or no battery at all when harvesting energy from the environment (e.g., solar, thermal). Such devices may be called low power.

Low power design which refers to the techniques used to minimize the energy consumption of a product and thereby increase energy efficiency [10]. Unlike cell phones, which require a fair amount of power but are easily recharged from a wall outlet anywhere, some IoT products must operate for days or months without access to traditional battery chargers. For example, an oil infrastructure IoT monitoring device in a remote field powered with solar panels. The solar panels do not produce energy at night, and if inclement weather reduces the output of the panels during the daytime, then the IoT monitoring product may not have access to additional power to charge its battery for weeks. Implementing methods of low power design can mitigate this kind of situation. Questions for low power design of IoT products include:

- When should the device enter a low power mode (also known as 'sleep mode')?
- When should the device wake up from sleep mode?
- What minimal functionality should the device maintain on entering the sleep mode?
- How long should a device be allowed to stay in sleep mode?
- Should a device in sleep mode retain memory of its state before it entered sleep mode?

Hardware and software design techniques exist for low power design [10] and both should be considered when starting a low power IoT project. With hardware, it is important to choose a microprocessor that fits operating design goals but also enables low power operation. In this context, the following three key aspects should be considered:

- Does the microprocessor operate off low voltages?
- Has a microprocessor been chosen that does not have extra unused hardware features or peripherals?
- Does the microprocessor support a microkernel that allows the efficacy for low power operations versus full power operations?

Some other choices for low power design include:

- Using integrated semiconductor sensors rather than discrete electronic sensors, can be generally lossy.
- Selecting the most power-efficient wireless protocol that meets the device's communication specifications. For example, if short range transfer of a small amount of data is sufficient, Bluetooth Low Energy (BLE) can be used. If longer range communication is required, cellular or satellite protocols can be used.
- Taking advantage of Power Management Integrated Circuits (PMIC), which employ specific control techniques to ensure efficient operation at appropriate voltage levels.

When considering embedded software in low power designs, one should be sure to plan for a sleep mode state at the system architecture level. Some processors may have a simple single command to enter sleep mode. However, if the time and place to enter sleep mode has not been thought through at the architecture level, then it can be very difficult to ensure the software manages the sleep mode state appropriately. At the design level, algorithms and code can be written and optimized for power efficiency. Optimizations could be intentional, such as minimizing resource-intensive computations in required calculations or carefully scheduling tasks to reduce power consumption. They could also be the result of using power-efficient coding languages or compiler directives in tandem with the appropriate microkernel configurations for low power output.

5.7 Design for Secure Field Deployments and Updates

It is important to maintain device and data security throughout the lifetime of IoT products. Many IoT products can simply be thought of as data acquisition systems. In some applications, the data collected may not be sensitive in nature. Examples include IoT weather stations, smart lighting, and robotic vacuums. In other cases, the data may be highly sensitive and could compromise personal privacy. For example, patient monitoring devices in healthcare, utility infrastructure applications, and facial recognition products deal with sensitive data. Even for IoT products whose data may not be considered sensitive, the data acquired can have value to both the consumer and manufacturer. For all these products, updates may be needed over time to improve functionality or address vulnerabilities. With IoT products that may have exposed hardware ports as well as embedded software stored and running on device hardware rather than in the Cloud, the task of securely installing IoT hardware and updating IoT embedded software becomes quite challenging.

There are many good practices in digital device security including anonymizing data, encrypting data, and reducing the exposure of transmitted data [11]. These and other practices should be considered and applied as appropriate by the IoT product team. However, the emphasis in this section is on IoT product deployment and embedded software updates. As with low power design, these aspects of IoT products should be considered early in the product development life cycle.

IoT products may need hardware updates or embedded software updates throughout their lifetime. Hardware updates always require hands-on work and can involve opening the product in the field to modify its electronics or it could require swapping out the entire IoT product with an updated design. Embedded software updates could be done in a hands-on fashion with code being loaded onto storage media (e.g., USB drive, SD card) and then plugged into the IoT product for update. They could also be done in a remote fashion over the Internet or a local network, eliminating the logistics challenge of obtaining physical access to all IoT products that need update.

While not typically needed as frequently as software updates, hardware updates should be considered for IoT products which may have long serviceable lifetimes or that deal with safety critical concerns. When considering the ability to upgrade hardware, three areas require special attention: the base hardware platform, modular design, and access to physical ports. The base hardware platform includes key components like the embedded microprocessor or microcontroller and the support circuitry necessary to enable functionality for those components. In any embedded project, it is important to specify the minimum processing power and to ensure that the peripherals included support all the necessary hardware. It can be attractive, and can reduce complexity, to choose a microcontroller that satisfies the minimum requirements of a product. However, designs based on this approach are unlikely to support hardware upgrades without replacing entire electronics boards. Instead, it may be worthwhile for a design team to brainstorm 1, 5, or even 10 years into the future to imagine what additional algorithms, processing power and memory may be needed for a particular product. If a microcontroller or platform exists with specifications that meet anticipated future needs without overly challenging the design team, it should be considered as an option to future-proof hardware upgrades.

Certain aspects of IoT products are more likely to evolve with time and need update than others—leading them to benefit from a modular design approach. A review from 2015 shows the numerous wireless communication protocols developed over time for home and neighborhood area networks including Wi-Fi, Zigbee, Bluetooth, and Cellular Networks. Since then, even more advanced networks and protocols such as Wi-Fi 6 have been developed to handle the increased integration of IoT products. With this knowledge, it may be prudent when designing an IoT product to enable the wireless technology through a generic hardware expansion interface. Many wireless transceivers are available as expansion boards, such as the Digilent Pmod ESP32 Wireless Communication Module [12]. The hardware designer can create a pluggable interface on the circuit board to support these types of cards and thus allow modification of the wireless technology used by the IoT product if necessary. This can be done as opposed, or in addition, to choosing a microcontroller with an integrated wireless module. The modular design approach provides another option to solving the challenge of future-proofing the base hardware platform. For example, if a distributed processing approach is enabled through the inclusion of expansion boards to the core platform, additional microprocessors, microcontrollers, or storage can be installed in the field to enable upgraded functionality.

It is also important to limit access to physical ports of any sensitive IoT products deployed in the field as physical attacks are commonly used to breach security [13]. Some data ports, such as USB ports, may be included in product and accessible under certain conditions by design. For example, a service field technician may need to connect a thumb drive to an IoT product to retrieve data or to update software. Other data ports may only reside on the internal circuitry with no intention of being used in the field. Microprocessor emulation ports like JTAG are commonly used to program and debug an embedded system during the design phase of a project. For both types of ports, access should be limited when the product is in the field. Ports intended to be used for service or maintenance should be made inaccessible behind enclosure covers that are locked or otherwise require specialized tools to open. For ports only used during the design phase of a project, consider removing the physical connector from the circuit board or potting them prior to release into the field.

Software updates are generally more common in products, and it is not difficult to find best practices to consider during design and when managing updates [14]. With IoT products, the embedded nature of the software does not disqualify it from applying those best practices. In this context, it is important to ensure that the embedded software has an on-board ability to apply techniques for secure software updates that the product team has chosen to use. This can include decryption of software packages, integrity, and digital signature verification. One option is to embed this functionality in a standalone bootloader application which resides in the embedded system and is called to execute when software updates are requested.

As with any design decision, trade-offs exist for the security considerations mentioned earlier. It is up to the project team and management to carefully consider the pros and cons of implementing security features that address deployment and field updates.

5.8 Production Electronics and Enclosure Hardware Design

In Sect. 5.4, we discuss prototyping the hardware. In this section, some considerations for production electronics and integration with hardware enclosures are covered. Since manufacturing and assembling circuit boards can take several weeks, the goal for the project team should be the need for only one production "spin" after all planned prototype "spins" have been completed. A "spin" typically refers to a round of circuit board design and manufacturing. To reduce the chances for additional production spins, it is important to consider aspects of integrating with the mechanical enclosure, electromagnetic shielding, and other electrical safety requirements that may be imposed based on the product classification [15].

Perhaps the most important question when integrating an electronics board with a mechanical enclosure is: does the board fit? Typically, an enclosure has standoffs to which electronics are attached via mounting holes on the PCB. The mechanical and electrical team must carefully coordinate the geometry of the circuit board and location of the mounting holes to ensure proper fit. In addition to board mounting, the panels of enclosure bodies may have cutouts for the electronic connectors. These cutouts may interface with board-mount connectors (connectors which are soldered directly onto PCBs) or panel-mount connectors (connectors which attach to the enclosure body and are connected to the PCB through a wire harness). As with mounting holes, the placement of connector interfaces must be coordinated between the mechanical and electrical teams. See Fig. 5.1 for a simple PCB enclosure example [15].

A consideration important for all electronics is electromagnetic shielding. Shielding protects the environment and other equipment from undesirable electromagnetic interference from a circuit. It also protects the circuit itself from interference outside its enclosure. Shielding becomes even more important for IoT products given they oftentimes include wireless transceivers. In these cases, electromagnetic shielding should be carefully designed to block unwanted emission and reception of unused wireless frequencies but to allow strong transmission of the IoT product's operating frequencies (e.g., 2.4 GHz for Bluetooth).

Regulations for IoT-based devices are mentioned in Chap. 9. In relation to production electronics, the design team should keep in mind all applicable safety

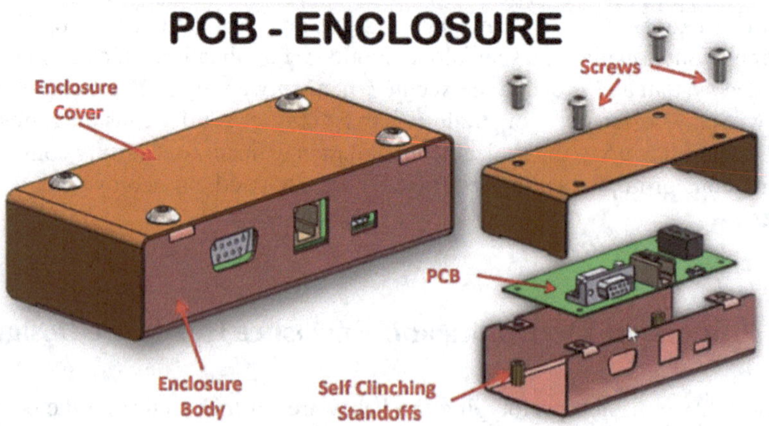

Fig. 5.1 PCB Enclosure showing mounting and connectors [15]

requirements based on the type and risk level of the product. There are several aspects of electrical design and mounting, such as creepage distances and clearances that need to be maintained in the final product to ensure safety compliance [16]. It is recommended that prior to manufacturing a production spin of electronics and enclosures, a careful review of electrical safety requirements.

5.9 Mobile or Web-Based Interface

IoT products will often have a simple user interface including buttons, switches, and a small screen to display information on the device. However, more advanced functionality can be exposed with a Cloud interface. This can take the form of a mobile phone or tablet application, or a Web-based application run through a browser on a computer. Based on the use cases for the product, the options for the development team are to deploy either (or both) of these options. The choice should be made after weighing in the pros and cons of each approach. Chapter 4 offers some insight highlighting the differences between Progressive Web Applications and Native mobile applications in Cloud projects. For IoT products, the anticipated user interaction with the device and the deployment location are some additional factors that can heavily impact the choice of the Cloud interface.

Consider a set of IoT cameras placed near doors and offices on the floors of an office building for security. After initial placement and setup, it is quite likely that there is little, or no interaction needed with the internet-connected cameras. The building manager or security team, who may reside in an office in the building or be located off-site, are the typical consumers of video footage. From a central location, they may want to monitor all camera feeds and react to embedded video processing such as motion detection or some untoward incident. In this case, a Web-based interface with a Cloud application is desirable. With additional Cloud processing power, advanced security features such as facial recognition and detection of undesirable activities could also be implemented.

Now consider a set of IoT cameras attached to trees in a wildlife refuge for monitoring the count and movement of the local fauna. Some of these cameras may not have a reliable Internet connection. It can be expected that a person will be needed to visit each camera to physically off-load images or to transfer data using a near field communication protocol to portable device like a cell phone or tablet. Once a Mobile application is used to acquire the images, they can be retransmitted over the Internet and stored online. Similarly, data can be stored online via a hub-based approach where cameras without Internet access relay their information to strategically placed cameras or nodes where Internet access is available. At this point, they can be viewed with any photo viewing software and so the team need not recreate this functionality with a Web Application. This sort of approach also mitigates the issue of poor connectivity or infrastructure in developing areas.

These examples emphasize how interactions between IoT products and users, even when the IoT hardware may have similar functionality, can drive the choice whether to develop a Mobile or Web-based application. Since the choice can significantly affect overall project timeline, it is important for the project manager to elicit these scenarios in the early stages of a project.

5.10 Cloud Connection

There are many options available to connect a device to the Cloud, enabling IoT. At a minimum, it is necessary to provide the required hardware connectivity modules such as Bluetooth, Wi-Fi, or Ethernet along with the software drivers to operate these modules. Other factors to consider are Cloud platform support and communication protocols.

Many companies offer tailored Cloud support for IoT. This can include connection management and secure communication between thousands of simultaneously connected devices. Amazon provides AWS IoT Core, Microsoft offers Azure IoT Hub, and IBM has Watson IoT. Each platform will have strengths and weaknesses and can be evaluated on a per project basis.

Whether or not a popular Cloud platform is used, or an organization chooses to develop and maintain its own IoT Cloud management, the standards used for the communication protocols should be agreed upon. Three common IoT data protocols are Message Queuing Telemetry Transport (MQTT), HyperText Transfer Protocol (HTTP) and Constrained Application Protocol (CoAP) [17]. MQTT is a lightweight protocol requiring very little bandwidth and includes three levels of service ranging from Level 0 (no guarantee of message delivery) to Level 2 (delivery is guaranteed without message duplicates). This allows for communication even when the network may be unreliable. HTTP is a well-known data protocol for the World Wide Web. Despite not being optimized for IoT products, its pervasiveness on the Internet and support available make it worthy of consideration. CoAP can be thought of as a hybrid protocol which supports both low bandwidths but is also structurally like HTTP in terms of the client/server communication.

5.11 Challenges in IoT Testing

IoT products present many challenges which can be daunting. This is because IoT products are a combination of digital and analog electronics, embedded and host-based software, network protocols, and communication all overseen within a regulatory framework. According to a 2024 study [18], the time to launch an IoT device grew by 80% from 2020 despite advances in IoT technology and market familiarity. This was attributed to four main reasons:

- Product and device complexity.
- Regulatory hurdles.

- Higher security scrutiny.
- More advanced use cases.

All the above reasons have the chance to significantly impact and increase project test timelines. A detailed and comprehensive test plan can be developed that targets some of the major challenges in IoT product design. Some potential mitigations are considered below for each of the categories.

Product and Device Complexity With improvements in sensor and processor technology, IoT products have grown from sampling and reporting simple sensors at slow frequencies (e.g., a room temperature sensor) to orchestrating a coordinated result from multiple sensors with millisecond deadlines (e.g., an integrated home environment product that senses temperature, humidity, light, noise level and activate an alarm if smoke is detected or a home security issue is detected). Some sensors on a single device may have the option for several different communication protocols, increasing the testing complexity manyfold. To address the interoperability of multiple sensors in a product, it is wise to choose a minimal set of standard communication protocols that meet the device's requirements and to design test suites that are focused on those standard protocols. Additionally, end-to-end test frameworks and simulation environments allow for increased coverage when dealing with complex IoT products and devices.

Regulatory Hurdles Different countries have different regulations for IoT products. These regulations cover device safety, data privacy, and allowable use of the electromagnetic spectrum. Each country may have their own specific requirements and all countries regularly update their requirements for the common good. Some regulatory compliance scenarios are mentioned in Chap. 9. The keys to success in dealing with regulatory hurdles are to be informed and to plan for the cost and schedule impact of compliance testing. Project teams should be aware of the requirements of the countries in which the IoT product is intended for sale. An analysis of those countries' regulations should be done throughout the product development cycle to ensure appropriate product requirements are developed and met. This analysis should also feed into early project estimates of cost and budget. For example, in the USA, regulatory compliance testing can cost tens of thousands of dollars and take anywhere from 6 to 8 weeks including completed test reports from third party test houses.

Higher Security Scrutiny Large-scale data breaches can have significant costs associated with them both in terms of customer trust and financial impact. In 2022, the average cost of a successful IoT cyber-attack was more than $330,000 [19]. With over 16 billion IoT products online at the end of 2023, hackers have focused on IoT and are taking advantage of their typical lack of security compared to laptops and workstations [20]. Verkanda IP cameras and Roku TV streaming devices are high-profile examples of IoT data breaches. These attacks have raised awareness and placed emphasis on IoT product manufacturers to carefully consider cybersecurity. As mentioned in Chap. 11, implementing strict Secure by Design principles and performing rigorous penetration testing (pentesting) throughout product development may reduce the probability of successful attacks. Early planning for these efforts by the project team will ensure cybersecurity is anything but an expensive afterthought.

More Advanced Use Cases Advances in sensing and actuation peripherals enable IoT device designers to provide more and more value to customers through advanced features. Where a simple dashboard displaying heart rates during exercise may have been sufficient for a fitness tracker, now customers can expect to see multiple metrics like heart rate, steps and route taken, pulse oximetry, and breath rates at all times of the day whether exercising or not. These consumer demands place extra effort on systems engineers and product designers to develop thoughtful use case scenarios that allow for comprehensive and robust feature testing. To avoid lengthy test cycles caused by poor requirements, it is best to develop use case scenarios with methods that have been proven over time. Written in 2000, Alistair Cockburn's Writing Effective Use Cases [21] is one of the many resources that could instruct readers on writing high-quality use case scenarios.

When DevOps are deployed as a mode of operation in project development, they support building solutions to all the above challenges. A key modality is Test-Driven Development (TDD) which emphasizes weaving together coding and testing [9]. With early and automated testing incorporated, IoT projects have an increased chance for success.

5.12 Summary

The IoT product development process has many facets that stand out from the typical software development process. However, both are similar in that they deal with complex products and businesses benefit by using Agile methodologies and responding to changes in the marketplace. IoT product teams should include engineers from several disciplines including systems, mechanical, software, and electronics with experience in analog and digital design. A DevOps team should be in place to support the entire project and facilitate TDD and API testing. All these teams should be able to adapt to innovations in hardware technology and changes in regulatory requirements. Staying abreast of the best IoT design practices, such as design for low power, can help development teams produce desirable products and minimize schedule and project risk. Focusing on the IoT product as an end-to-end system with Cloud software when testing is planned and performed is also beneficial.

Appendix: IoT Development Platforms

To increase the market share in what is expected to grow upward from the 16.6 billion connected devices seen in 2024 [22], manufacturers of hardware microcontrollers, sensors, and wireless modules are competing to provide easy to use development platforms (also known as evaluation boards) for designers. These platforms provide engineering teams with systems ready to use out-of-the-box to reduce the time to market. Some evaluation boards are even suitable to be used as modules in larger product designs.

A list of popular development platforms at the time of writing is provided below. Links are provided for additional details.

Processors/Microcontrollers/Single Board Computers

1. Raspberry Pi 4 Model B Arduino Uno REV3
   https://www.raspberrypi.com/products/raspberry-pi-4-model-b/specifications
2. Arduino Uno REV3
   https://store-usa.arduino.cc/products/arduino-uno-rev3
3. ESP32-Gateway
   https://www.olimex.com/Products/IoT/ESP32/ESP32-GATEWAY/open-source-hardware

Sensors/Actuators

4. Analog Devices SILENTSTEPSTICK Stepper Motor Driver
   https://www.analog.com/en/resources/evaluation-hardware-and-software/evaluation-boards-kits/silentstepstick.html
5. Adafruit HC-SR04 Ultrasonic Sonar Distance Sensor
   https://www.adafruit.com/product/3942
6. Pimroni Ltd. BME680 Air Quality, Temperature, Pressure and Humidity Sensor
   https://shop.pimoroni.com/products/bme680-breakout

Wireless Modules

7. Silicon Labs BGM220-EK4314A Bluetooth Module Explorer Kit
   https://www.silabs.com/development-tools/wireless/bluetooth/bgm220-explorer-kit
8. Silicon Labs SiWx917-DK2605A WiFi 6 and Bluetooth LE Dev Kit
   https://www.silabs.com/development-tools/wireless/wi-fi/siwx917-dk2605a-wifi-6-bluetooth-le-soc-dev-kit
9. ST Microelectronics STEVAL-FKI868V2 Sub-1GHz (860–940 MHz) Transceiver Development Kit
   https://www.st.com/en/evaluation-tools/steval-fki868v2.html

Points to Ponder

1. Why is it especially important to carefully capture meaningful requirements during the life cycle of an IoT product?
2. What are some of the differences between active and passive sensors? Also, identify the similarities.
3. When should a custom-printed circuit board or evaluation kit be used for hardware circuit prototyping?
4. What is "sleep mode" for IoT products?
5. Why should one use modular design in IoT products?
6. What should electrical and mechanical engineering teams consider when developing enclosures for IoT products?
7. What are the main reasons that have increased challenges to IoT product testing?
8. What are some critical work products for DevOps teams in IoT projects?

Answers

1. Why is it especially important to carefully capture meaningful requirements during the life cycle of an IoT product?
   (a) IoT products may include electronics, embedded software, communications, and mechanical subsystems. As such, there can be a high cost to iterating hardware designs during the development cycle. Fewer iterations can save a project time and money.
2. What are some differences between active and passive sensors? What are some similarities?
   (a) Active sensors require power to operate. They may need additional power circuitry and may transmit signals to gather data. Passive sensors operate by measuring or detecting natural energy and can work without power aside from a sensing circuit.
   (b) Passive and active sensors both measure aspects of their environment. They also are both typically used in conjunction with Analog-to-Digital Converter (ADC) integrated circuit chips.
3. When should a custom-printed circuit board or evaluation kit be used for hardware circuit prototyping?
   (a) Very high frequency digital circuits or circuits too large for hand assembly should use a custom circuit board for prototyping. Complicated integrated circuits may also benefit from custom circuit board or manufacturer provided evaluation boards for prototyping.
4. What is "sleep mode" for IoT products?
   (a) "Sleep mode" is another term for a low power mode which utilizes design techniques to minimize the energy consumption of a product and thereby increase energy efficiency.

5. Why should one use modular design in IoT products?
   (a) A modular design allows for IoT products to be re-configured as technologies evolve in fields that experience rapid innovation such as wireless technologies.
   (b) Modular hardware also allows for future-proofing of products by ensuring hardware can be upgraded in the field as software algorithm and processing needs increase.
6. What should electrical and mechanical engineering teams consider when developing enclosures for IoT products?
   (a) Printed circuit board fit and enclosure panel access should be considered to ensure smooth integration. For IoT products that rely on wireless communication, electromagnetic shielding is also important to consider in enclosure design.
7. What are the main reasons that have increased challenges to IoT product testing?
   (a) Product complexity, regulatory hurdles, higher security scrutiny, and more advanced use cases in IoT products present increased challenges to testing.
8. What are some critical work products for DevOps teams in IoT projects?
   (a) IoT projects follow a classic paradigm where DevOps is essential for success. Ensuring efficient workflows for embedded code development on microprocessors and FPGA devices is important. Similar workflows can be applied to electronics hardware development. DevOps teams should also facilitate TDD and other forms of automated testing.

References

1. Nadeem, M. A., Hasnain, M., Saleemi, M. M., Mohsin, M. A., Ansari, M. A., & Essalah, W. (2024). Challenges in requirements engineering for IoT solutions. *Journal of Computing & Biomedical Informatics, 6*(02), 55–63. https://jcbi.org/index.php/Main/article/view/333
2. https://www.neurasignal.com/product/novaguide-platform
3. Kamina, T., Aotani, T., Masuhara, H., & Tamai, T. (2014). Context-oriented software engineering: A modularity vision. In *Proceedings of the 13th International Conference on Modularity* (pp. 85–98). ACM.
4. https://www.amd.com/en/products/software/adaptive-socs-and-fpgas/vivado.html
5. https://www.intel.com/content/www/us/en/collections/products/fpga/software/downloads.html
6. Taraate, V. (2014). Design guidelines for FPGA based design. *International Journal of Engineering Research & Technology, 3*(11).
7. https://www.intel.com/content/www/us/en/docs/programmable/683323/18-1/recommended-design-practices.html
8. https://barrgroup.com/blog/perils-preemptive-multitasking
9. Bhatt, P. C. P. (2021). *Software design, architecture and engineering: Concepts and practice.* Prentice Hall India, Easter Economy Edition. ISBN-978-81-951611-4-0.

10. https://runtimerec.com/low-power-design/
11. https://learn.microsoft.com/en-us/azure/security/fundamentals/data-encryption-best-practices
12. https://digilent.com/reference/pmod/pmodesp32/reference-manual
13. https://deviceauthority.com/10-common-iot-hacks/
14. https://www.pureversity.com/blog/manage-software-updates
15. https://morepcb.com/pcb-enclosures/
16. Scopelliti, D. P. *Clearance and creepage*. D-Scope-Tech, Inc.
17. https://www.allaboutcircuits.com/technical-articles/internet-of-things-communication-protocols-iot-data-protocols/
18. https://iot-analytics.com/challenges-iot-product-launches-why-time-to-market-has-increased-80-percent-in-4-years/
19. https://www.psacertified.org/app/uploads/2022/08/Confidence_to_Create_Advisory_Paper_PSA_Certified-1-compressed.pdf
20. https://asimily.com/blog/the-top-internet-of-things-iot-cybersecurity-breaches-in-2024/
21. Cockburn, A. (2000). *Writing effective use cases* (1st ed.). Addison-Wesley Longman Publishing Co., Inc.
22. https://iot-analytics.com/number-connected-iot-devices/

Chapter 6
Planning Layout for the Project

6.1 Introduction

Project layout identifies the arrangement in time and space of the tasks and resources to execute a project. The project plan includes the way this layout is distributed across geographies. This is the primary rationale behind collecting requirements before starting to implement a project. However, as we described in the previous chapter, the user requirements may change during the execution of a project. It was for this reason that methods like Agile and DevOps have found wider acceptance. These methods are equally applicable for a Cloud software project. However, sometimes these may not be sufficient. The reasons may be cost overruns and confusion among the geographically dispersed teams which could occur when proper tracking methods and communication tools are not used.

In this chapter, we will start by explaining a strategy which begins with the end in mind, an idea borrowed from one of Steven Covey's seven habits of effective people [1]. Which means, we need to know two things: (a) Cloud services to deploy the application, and (b) costs based on expected usage patterns. Next, we will present ways to collect user needs and how to convert them into the software requirements. Converting user needs into system workflows and describing them in a precise language is an art in itself practiced by the software architects. They articulate system workflows to help the software engineers to finally write code.

6.2 Cloud Project Strategy

In general, a strategy is a plan of action or policy designed to achieve an overall objective, which in our case is timely completion of project within the allocated budget. In other words, strategy includes an understanding of customers (user

P. C. P. Bhatt et al., *Project Management in Cloud and IoT Applications*,
https://doi.org/10.1007/978-3-031-99427-2_6

personas), product architecture, technology stack (OS, Web platforms, languages, frameworks, and tools) and above all, getting the right implementation team in place.

One begins by understanding user needs, their profiles or personas and pain points, etc. The overall objectives are enumerated to highlight who will all use this product and why it may be the preferred option for them. This also entails a market analysis of competing products and services. Users generally pick a product that is better, faster, or cheaper than the existing solutions. If the proposed product can meet at least 2 out of 3 distinct criteria, then we may have a winner at hand.

A software strategy needs to outline the vision, goal, and high-level ideas of what will be delivered. A Cloud offering must also include elements of performance, reliability, and scalability. Performance refers to the response time to user actions accounting for end-to-end latency caused by networks and resources located in remote datacenters. It needs to be specific, e.g., that 95% of user queries will be responded to within 1 s. The performance in the Cloud can be improved by using additional resources and load balancers [2].

Reliability refers to uptime and availability. Although, a typical Cloud vendor may advertise their reliability as three 9 s (e.g., 99.9%), the solution architect will need to measure end-to-end reliability including networks, Wi-Fi routers, and any other services that may cause an interruption or delays. The reliability of operations can be improved by exploiting feasible concurrency and backup systems distributed across geographies although that will also increase the deployment costs.

Finally, scalability takes into account how a system will behave when number of users and their activities increase, i.e., can the performance be maintained? There are several ways to improve scalability in the Cloud [2].

As shown in Fig. 6.1, there are six pillars needed to support a successful Cloud-based application: Customers, Team, Products, Processes, PoCs, and Cloud infrastructure. We have studied PoCs, a product's development and test processes in an earlier chapter. In the next section, we will study how to tease out requirements from the customers.

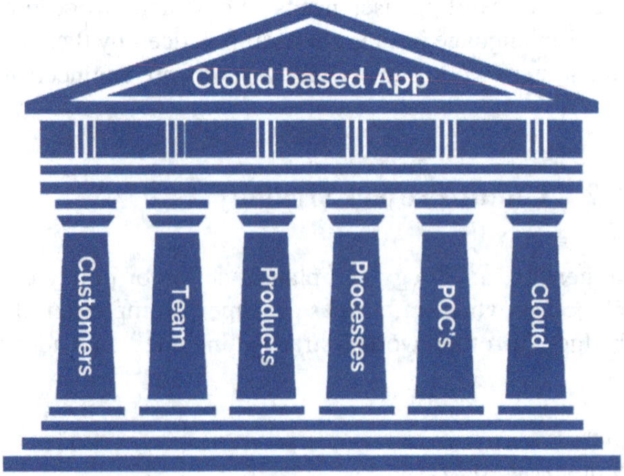

Fig. 6.1 Pillars to support a Cloud-based App

6.3 Art and Science of New Features

In a simple world, it would be sufficient to do a user survey, gather their inputs, and list out product's requirements. However, in a real-world scenario, there is more than one category of users. For example, for a medical device software application to telemedicine, it was observed that there are at least seven types of users. The intent of this project was to copy medical data from a brain ultrasound machine to a Cloud, so physicians can view it remotely. Given this context, one can easily identify the following seven user personas:

1. Reviewing Physician
2. Sonographer
3. Referring Physician
4. Hospital IT Manager
5. Clinical Researcher
6. Hospital Researcher
7. Patient

Once these personas and their user needs were identified, then knowing their pain points a new product's requirements can be captured as follows:

Reviewing Physician A *Reviewing Physician* is trained to interpret a patient's ultrasound examinations.

The Reviewing Physician is the key individual who reviews a brain scan, interprets the data, and signs the final examination report. Their typical examination reading workflow consists of the following:

(a) Review Patient Medical History/Context, within an EMR (Electronic Medical Records).
(b) Receive an alert that the report is complete, possibly at the end of the day if the report is uploaded to a PACS (Picture Archiving and Communications System) on a hospital via a wired or wireless connection.
(c) Access the report to review its findings.
(d) If warranted, explore the raw images captured by the Ultrasound Technologist stored in a PACS from the VNA (Vendor Neutral Archive).
(e) If warranted, may additionally label and add annotations to the images.
(f) Dictate notes or key in the examination report into the system.
(g) Digitally sign the completed report.
(h) If the findings are critical, then call the Referring Physician to review findings and plan next steps of care.

Following are the challenges and pain points encountered by the Reviewing Physician:

- Inability to hear the scan's sounds because audio is not captured in the current reports.
- Inability to see the moving images of a scan because it is stored in EMR, which does not support video formats.

- Inability to easily download video from medical device to their remote viewing station.
- Inability to review visual/audio raw data if data was deleted from medical device due to storage constraints.
- Inability to remotely monitor in real-time, as medical device is not connected to Internet.
- Potential lag between when a medical scan report is completed and when it is made available.

To address these pain points, a Cloud App should provide the following:

- Access to visual/audio scans away from the medical device.
- Support annotation and digital signatures from a remote location.
- Interface to the PACS to send edited images to the patient's folder. Also, interface with the EMR to send the final report.
- Redo scan's measurements and update report if device data was inconsistent.
- Protect patient data.

Additionally, the medical system should support:

- Connectivity to PACS in an environment with electronics noise.
- Near-real-time raw data streaming to a PACS and/or Cloud on critical scan measurements.
- Export complete scan data to the Cloud app.
- Protect patient data in transit.

Sonographer A *Sonographer or Ultrasound Technologist* is the individual collecting raw ultrasound data from the medical device.

A sonographer is trained to use the device, complete the scan, and prepare a medical report. Their typical workflow is as follows:

(a) Download the Modality Work List (MWL) from the MWL PACS server.
(b) Review the requested examinations and patient information for accuracy.
(c) Physically locate themselves next to the patient.
(d) Using a scan to collect the clinical data, in "unrefined format," from the patient during the examination.
(e) After the examination, transfer the raw ultrasound data to the PACS via USB or via a wired network connection.
(f) From the work computer, access review software on the VNA interfaced to the PACS.
(g) Using review software to ingest the unfiltered ultrasound file from the device because it does not have a full data (including sound) exporting capability.
(h) Use the review software to refine and filter the data.
(i) Use the review software to create the scan examination report.
(j) Send the examination report and images to the Reviewing Physician for review and approval.

Note: In most cases, the technologist does not see the report again after it is sent to the Reviewing Physician. After the Reviewing Physician changes/signs the report, it is automatically sent to the patient's imaging results record in the EMR so the treating physician can review the result.

The following are the challenges and pain points encountered by the Ultrasound Technologist:

- Transfer the data from the device with a USB.
- Connect the device via Ethernet to the local PACS server.
- Transfer device data into a software review and reporting program.
- Inability to create rapid reports on a higher priority (more critical) patient.
- Constrained by 1:1 nature of conducting a scan for each patient.
- Inability to create a more efficient workflow to perform the scans automatically on a current patient, while the technologist is completing reports on a previous patient.

To address these pain points, the Cloud App should provide the following:

- Access to visual/audio medical device scan apart from the scanner.
- Support remote viewing and listening visual/audio medical device scan data away from the medical device system for telemedicine.
- Connect to an EMR system via an API where a link to the final report and supplementary data associated with the patient could be provided, if appropriate.
- Connect to a PACS for storing examination images and data if the hospital requires this.
- Provide an ultrasound structure reporting export capability.
- Reproduce medical device findings, including the medical device measurements that inform the priority alert level in a PACS system.
- Protect patient data.

Additionally, the medical device system should support:

- Wireless connectivity to a PACS in an environment with electronics noise.
- Near-real-time raw data streaming to a PACS and/or Cloud on critical medical device measurements.
- Export complete medical device data to Cloud for consumption by the Cloud app.
- Protect patient data in transit.

Referring Physician A *Referring Physician or treating physician* is a physician who furnishes a consultation or treats a beneficiary (patient) for a specific medical problem and who uses the results of a diagnostic test in the management of the beneficiary's specific medical problem.

The Referring Physician seeks a medical device examination on a patient. Their typical workflow consists of the following:

(a) Review Patient Medical History/Context, within an EMR.
(b) Evaluate patient to determine if a medical device examination is needed.
(c) Annotate patient record in EMR explaining the rationale for medical device examination.

(d) Request a medical device examination, within an EMR.
(e) Read the final report documented by the Reviewing Physician.
(f) If needed, communicate with the Reviewing Physician to determine the next steps.
(g) If the findings are critical then the Referring Physician will receive communication from the Reviewing Physician or performing technologist to determine immediate next steps.
(h) Follow-up with patient on findings and next steps of care.

The following are the challenges and pain points encountered by the Referring Physician:

- Potential lag time in notification of critical diagnosis or change in patient's status as a consequence of how the medical device system is connected to the PACS. To address these pain points, the Cloud App should provide the following:

- Interface to an EMR system.
- Interface to the PACS.
- Reproduce medical device findings.
- Enable real-time conferrals on critical findings by the Reviewing Physician.
- Protect patient data.

Additionally, the medical device system should support:

- Seamless wireless connectivity to the PACS/VNA/Cloud for uploading reports.
- Near-real-time raw data streaming to a VNA and/or Cloud on critical findings by the medical device.
- Export complete medical device data to Cloud for consumption by the Cloud app.
- Protect patient data in transit.

Additionally, the Cloud App must include and support the following individual roles and functions:

Hospital IT Administrator A *Hospital IT Administrator* is the individual responsible for ensuring that vendor systems are securely and safely integrated into the hospital IT ecosystem. The hospital IT administrator is generally responsible for ensuring that vendors comply with the following regulatory and compliance requirements of the hospital:

(a) Secure and protect patient data in use, in transit, or in storage.
(b) Establish access controls.
(c) Audit and track who accessed the systems and what was accessed.

The following are the challenges and pain points encountered by the hospital IT administrator:

- Install Ethernet ports or turning on Ethernet ports in rooms where a wired-only medical device system might be used.
- Troubleshoot vendor devices within their IT ecosystem or calling vendor for support.

- Buy additional storage capacity for storing large imaging studies for each patient.
- Buy more storage capacity for archiving imaging studies over the lifetime of the patient (or for 40 years).
- Facilitate remote viewing of medical device images to physicians—especially to those physicians on call.
- Facilitate remote viewing of medical device video to physicians—especially to those physicians on call.
- Ensure that device-generated data goes through an internal server integrated into the hospital infrastructure before that data is potentially uploaded into a Cloud system.

The Cloud App should provide the following:

- Connect to the hospital EMR system through an EMR API.
- Connect to the hospital PACS system to upload "golden" report if required by the hospital.
- Support Active Directory (e.g., LDAP) to manage personnel access rights.
- Enable Reviewing Physicians that are on call to do remote viewing of medical device data.

Additionally, the medical device system should support:

- Connectivity to an EMR/PACS in an environment with electronics noise.
- Connect to a VM (virtual machine)/server within the hospital's firewall before any data is forwarded to Cloud.
- Software updates compliant to the Hospital's established requirements.

Clinical Researcher A *Clinical Researcher* is responsible for developing the "intended for use" of the medical device system application and its data.

A typical workflow for the clinical researcher is as follows:

(a) Establish or reinforce a research hypothesis.
(b) Identify data sources, methods, devices, and environment to test hypothesis as part of a clinical research trial(s).
(c) Establish a protocol for conducting clinical research that addresses patient safety, patient data protection and security, ethics, informed consent, and additional safeguards.
(d) Collaborate with other clinical researchers.
(e) Obtain IRB (Internal Review Board in hospital) approval for the protocol which includes the details of the clinical research trial.
(f) Sign a BAA (Business Associate Agreement) if required by external entities employing the clinical researchers.
(g) Execute the approved research protocol.
(h) Collect data, Analyzing Data, Visualizing Data.
(i) Report and publish the results.
(j) Obtain IFU (Indications for Use) by the FDA (Food and Drug Administration), if warranted.

The following are the challenges and pain points encountered by the clinical researcher:

- Delayed availability of the medical device data for the clinical trial because whenever the data is manually uploaded to a USB and subsequently transferred to protected data repository for later analysis.
- Gather additional patient data to support the analysis of medical device data.
- Gather other device data with labels to support a comparative analysis of medical device data.
- Not being able to access system logs to enhance their analysis of device induced variation within the captured signals.
- Not being able to share data analysis and visualization with external clinical researchers separated from the medical device system.

The Cloud App team should develop a product to support the following:

- Connect to the hospital EMR system through an EMR API.
- Upload from the EMR relevant patient data as established by the IRB and the BAA.
- Connect to the hospital's VM/server for uploading medical device data if so required.
- Present visualization of data and data analysis.
- Present summary of data analysis outcomes.

Additionally, the medical device system should support:

- Connectivity to the EMR/PACS in an environment with electronics noise.
- Sending data to Cloud directly or indirectly via the VM/server within the hospital's firewall.
- Sending system logs to medical devices directly or via the VM/server in the hospital.

Hospital Researcher A *Hospital Researcher* is responsible for conducting clinical trials in collaboration with Clinical Researchers.

A typical workflow for the hospital researcher is as follows:

(a) Collaborate with fellow clinical researchers.
(b) Obtain IRB approval for the protocol which includes the details of the clinical research trial.
(c) Enter into a BAA when needed.
(d) Execute the approved research protocol.
(e) Label the data.
(f) Report and publish results.

The following are the challenges and pain points encountered by the hospital researcher:

- Lacking immediate availability of the medical device data as part of the clinical trial because currently data must be manually downloaded into a USB and then transferred to a protected data repository for later analysis.

- Obtain additional patient data to support the analysis of medical device data.
- Obtain other device data, with labels, to support a comparative analysis of medical device data.
- Not being able to access the data analysis and visualization, separated from the medical device system.

The Cloud App should support the following:

- Connect to the hospital EMR system through an EMR API.
- Upload from the EMR relevant patient data, as established by the IRB and the BAA.
- Connect to the hospital's VM/server for uploading medical device data if so required.
- Present visualization of data and data analysis.
- Present summary of data analysis outcomes.

Additionally, the medical device system should support the following:

- Connectivity to an EMR/PACS in an environment with electronics noise.
- Send data to Cloud via the VM/server within the hospital's firewall.

Our description so far has been for a Cloud-based Application for medical devices. However, a general healthcare service application should also meet needs and expectations of a patient. Therefore, it is important to understand a patient's concerns and the likely workflow. For more details and general healthcare flow, see [2]. Later in the book, we will discuss the outline of a Cloud application dedicated to Patient's needs.

Patient A *Patient* is a person seeking healthcare services. A patient may either have a disease or suspects to have one needing a diagnostics test. Once the test results are available, a discussion with the Referring Physician may be required to interpret and explain the results. In some cases, a patient may want to take his/her report to another physician for a second opinion.

A typical workflow for a patient is as follows:

(a) Approach a Referring Physician seeking medical care, who may order one or more diagnostic tests.
(b) Get the tests done.
(c) When the report is ready, meet with the physician to understand the results and for next set of steps.
(d) In some cases, ask for a copy of the report to share with another physician for a second opinion.
(e) During treatment to continue meeting with the physician.
(f) Do additional tests as recommended by the physician.

Following are the challenges and pain points encountered by the patient:

- Lack of timely medical care if the physician or testing facilities are busy or not available.

- Ascertain the correctness of results.
- Need reassurance that proper care is being given.
- Ensure that medical billing is correct, and insurance company is covering the cost.

The Cloud App should provide the following:

- Connect to hospital's appointment system to schedule medical tests.
- Present visualization of data and data analysis.
- After a test is done, send the report to hospital's EMR systems.

Additionally, the medical device system should support the following:

- A mechanism to download the results files so these can be shared with the patient.

In this section, we comprehensively enumerated the requirements of different stakeholders. Next we will examine the Cloud project's resource planning.

6.4 Cloud Project Resource Planning

Once the requirements for a new Cloud-based App are understood and documented in a clear, unambiguous, and non-contradictory manner then the next order of business is to estimate the resources needed to execute this project. Also, some PoCs may be needed to establish feasibility of new technologies so that a real-time streaming solution can be deployed in the public (or private) Cloud. This solution can be used to remotely monitor a patient's examination or a mission-critical notification of an alert sent from a medical device.

Instead of putting all users' requirements in a single Cloud app's first release, it may be prudent to plan time ordered and graded releases of the product. These are summarized in Fig. 6.2, as three key requirements from a physician's perspective. Such a roadmap gives an opportunity to introduce a base product in a matter of months to get users' feedback while working on the next release. Also, this will surface potential barriers to adoption. This may help in alleviating issues such as hospital's IT security concerns for sending medical device data to Cloud. Additionally, it will help to work through the deployment process of the current version, while the development team starts to work on the next version.

The cost should be estimated to account for the development and test teams' salaries and other resources required such as compute time, storage, and networking costs in the Cloud and compensation for users to get their feedback on intermediate versions of the product.

After the requirements from all the user types are captured, the next step is to model the solution. That is how a language called "Unified Modelling Language" (UML) came about. A model can be generated using UML or Petri net or its extensions, such as Morphic net [2].

UML is presently one of the well-accepted frameworks in creating object-oriented models. A detailed discussion of UML is beyond the scope of this book. It

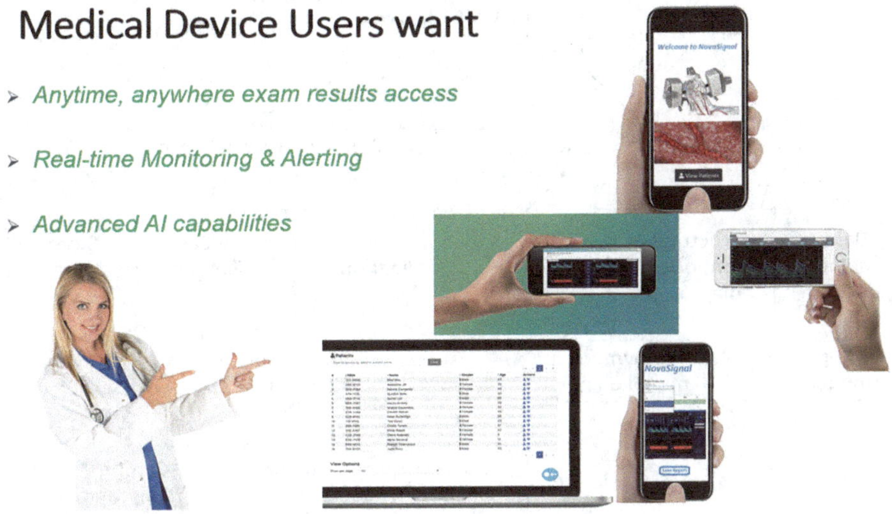

Fig. 6.2 Keys requirements from physicians for a telemedicine App

can be found in [2]. Here we describe the minimal essential details so that the basic process of modeling can be understood.

The visual UML descriptors form a basis to evolve system architecture and design, as shown in Fig. 6.3. It should be remarked here that UML description of a system is very useful for developing TDD and BDD tests as well.

Let us first describe some key UML capabilities:

1. A framework which helps in visualizing interaction between entities. This helps in ensuring that all the operational scenarios are captured explicitly. So, to that extent it is complete.
2. UML class definitions identify all the associations and assure fulfillment of contractual roles for every defined entity.
3. The UML depiction leads to robust documentation with no ambiguities. Part of the strength of UML is the way various UML diagrams are created and drawn.
4. A UML diagram offers both structural and behavioral description. These two descriptions are complementary in nature to fully model a system.

Now we offer a brief UML description of the app's operations from the perspective of Reviewing Physician. Our focus shall be to capture the minimal structural and behavioral aspects. The structural aspects are captured in Fig. 6.4 using a class diagram. As is evident from the requirements, the Reviewing Physician has associations with the Ultrasound Technologist and the Referring Physician who have their own attributes and methods. Though technically, there may be multiple review physicians—for simplicity we depict the app's structure as if there is one Reviewing Physician who interacts with an Ultrasound Technologist, and a Referring Physician. This would depend upon a patient's condition. The point that needs to be understood

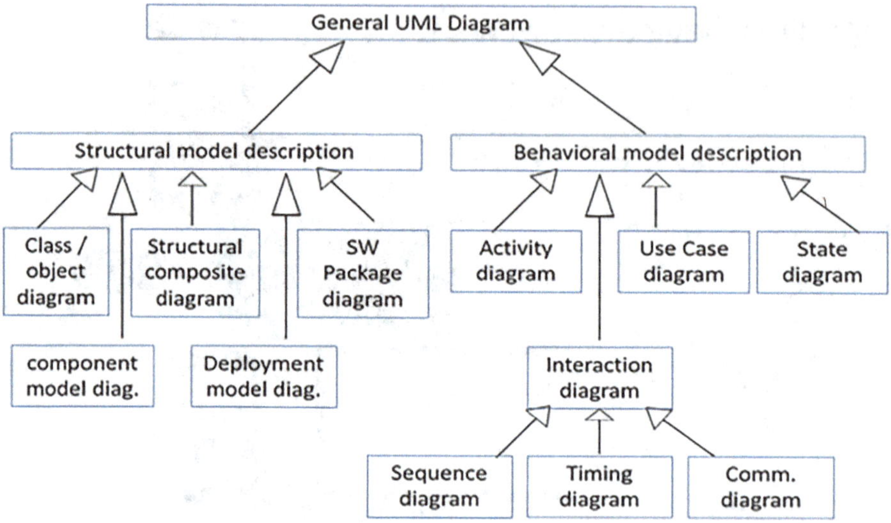

Fig. 6.3 Some of the key UML diagrams

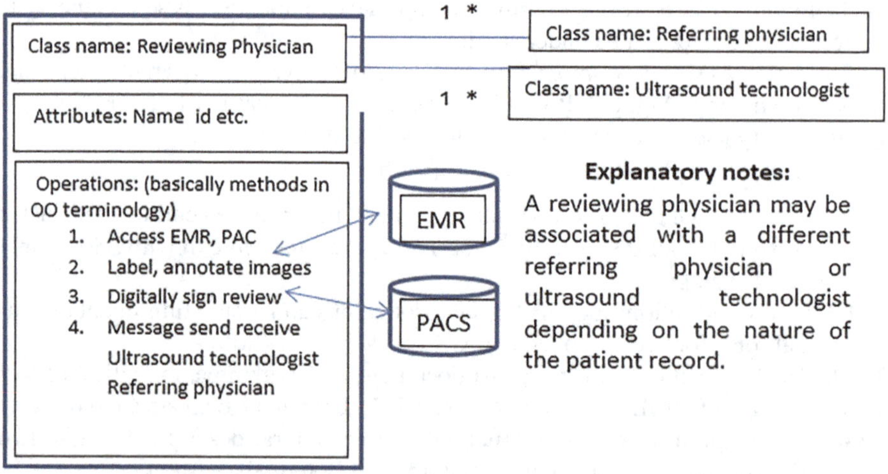

Fig. 6.4 A structural diagram showing Cloud App's operations for a Reviewing Physician

is structurally how the App may be designed with the identification of classes to model the entities that are inherent to the functioning of the system. The operations of labeling and annotating images to PACS and the uploading of report to EMR are not explicitly shown in Fig. 6.4 but can be easily inferred.

Next let us visualize the behavioral aspect for the entity 'Reviewing Physician'. This is shown in Fig. 6.5 as an activity diagram drawn using UML notation.

As one can see in Fig. 6.5, there are three swim lanes—one each for the Reviewing Physician, the Ultrasound Technologist, and the Referring Physician. The diagram

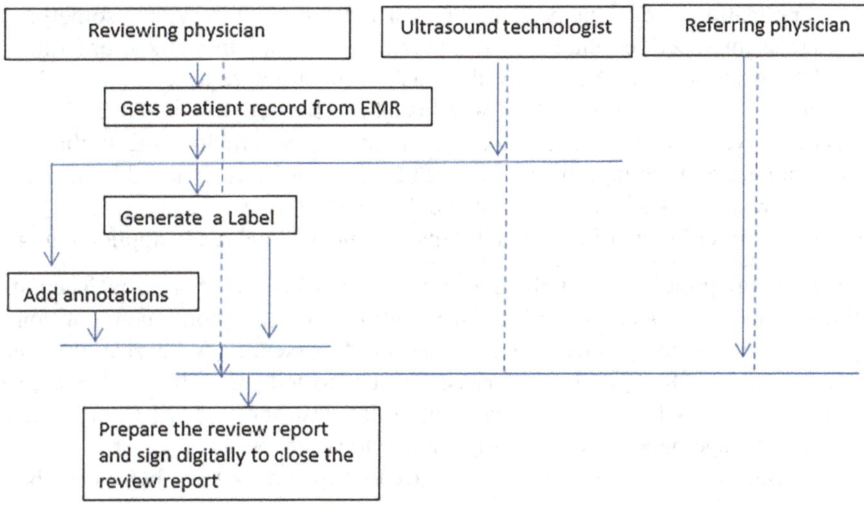

Fig. 6.5 An activity diagram showing behavioral aspects for a Reviewing Physician

shows the activities which may be warranted like seeking the Ultrasound Technologist's inputs before labeling and annotating. Once that is accomplished the review physician may seek the input from Referring Physician before digitally signing and closing the review.

6.5 Cloud Project Execution Frameworks

Once a roadmap, a SW architecture and development plans are ready, then the next steps are to estimate time and assemble a team with necessary skill sets, and other resources required to implement the Cloud-based products. This can be done through PoCs and project team's previous experiences. It is better to breakdown tasks in smaller sub-tasks to develop each one iteratively as outlined in the Agile [2] section of the previous chapter.

The advantages of developing software products in Cloud are the availability of ready-made frameworks and many open-source libraries to kick-start development. These can help in accelerating product development. For example, a team can use existing development tools, middleware technologies, and database management software to help with the creation and control of Cloud-based applications.

Let us again consider the medical device application discussed in the previous section. Suppose we were to store an mp4 file in a container then several ready-made streaming solutions are available to play it back with the controls to pause, forward, or reverse the video frames. The ready-made solutions save writing new code, and also offer optimized and well-tested libraries to quickly put together all related applications.

The application itself can be defined as a collection of services, sometimes referred to as microservices that are stitched together to implement the user requirements. These services can be deployed at IaaS (Infrastructure as a Service), PaaS (Platform as a Service), or SaaS (Software as a Service) tiers of the Cloud [3]. The application development team can then quickly model and architect the application for performance and scaling. The security angle has to be borne in mind for the data in transit, when the data is being stored and during data usage.

Below we describe the five essential steps to build a Cloud ready application [4]:

1. **Design:** An application is defined as a collection of services or APIs. The data flows from one service to another. These services are then combined to accomplish a complete composite application. This is the essence of a Service-Oriented Architecture (SOA) [3]. Each service can be hosted on a different machine instance, such as MYSQL for searching a database, or an IAM (Identity and Access Management) service for login authentication, etc.
2. **Decouple:** A Cloud represents a complex distributed system that works best with application architecture which breaks out processing and data into separate components. Once the data is decoupled from the application, it can be stored or processed on multiple public or private Cloud instances. For example, identified PHI (Private Health Information) can remain in a separate storage in a private Cloud bucket, while the de-identified data for patients is sent to public Cloud for processing. Then the results are returned to the hospital for re-identifying the data and presenting them to the physicians.
3. **Communications:** Simply decoupling data and services by itself is not sufficient for properly architecting a Cloud-based application. Expansive application components that frequently communicate may lower performance and increase the cost of an application deployment whenever there is very frequent network I/O (Input-output). Therefore, the design of the application needs to optimize communication between the distributed components not running on the same platform.
4. **Model:** The definition of a performance model is critical for it to scale under varying load conditions. The team needs to study how the application will behave if the number of users increases by several orders of magnitude. This represents increased stress on the database servers, computational elements, and networking infrastructure. One needs to deploy monitoring tools to continuously measure an application's latency experienced by the users. Some services such as AWS (Amazon Web Services) offer tools like Cloud Watch [5]. It offers monitoring and observability services to DevOps engineers on the CPU utilization of Cloud instances, triggers alarms for actions whenever pre-defined thresholds are reached to start additional instances or withdraw some resources. The measurements are logged for debugging and tracking usage patterns of the Application users, as shown in Fig. 6.6. It shows the process of collecting metrics, real-time monitoring and sending alerts if necessary.
5. **Security**: For the applications hosted in a public Cloud, security considerations should be a high priority and not an afterthought. Basically, security needs to be

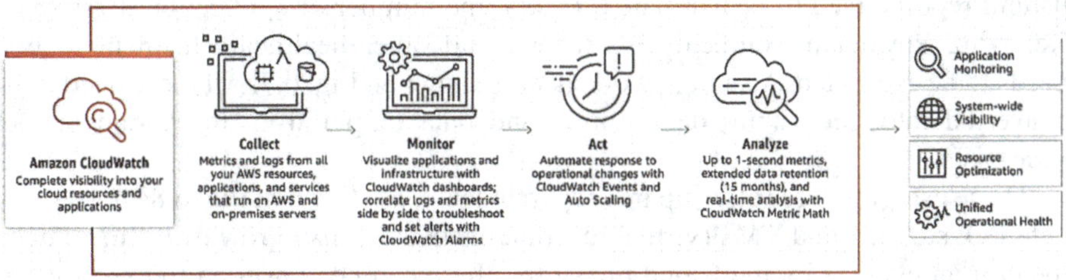

Fig. 6.6 AWS Cloud Watch tool [5]

built into the application's architecture. Healthcare data requires additional security under HIPPA [6]. For our medical device under consideration, security may begin with encrypting the data inside a hospital's firewall on the device and before sending it to the Cloud using hashed keys. Alternatively, one may track transactions by a blockchain. This prevents a man-in-the-middle attack [3]. This secures the data in transit over public nodes. Any attempt to tamper will result in altering hash values. Furthermore, data can be stored as encrypted in Cloud storage instances and the application can decipher it during run-time as needed. Note that this may affect the run-time performance of the application in the Cloud. Some Cloud service providers, such as Google offer native encryption at the storage levels [7].

Let us now focus on project budgeting. We need to figure out how a given business plan can be executed by estimating the costs involved.

6.6 Cloud Project Budgeting

A new product must fit in the business plan of the organization and needs to be priced for success. Pricing is determined by cost and the desired gross margin. An interesting fall out and benefit of the Cloud deployment costs is that one pays relatively less in the initial phase of the operation when the user base is low. This may allow companies with smaller budgets to release a project and plan to scale as the customer base expands.

For Cloud-based application projects, costs can be divided as NRE (non-recurring expenses) and recurring expenses. While the former refers to development costs, which are one-time in nature, the latter includes maintenance and deployment costs. The salaries of the team members, other overheads such as healthcare insurance and building costs may be easily computable. However, it is tricky to compute the Cloud deployment costs as these depend on the number of medical devices installed per quarter in the hospitals and the number of patient examinations done every year. These data points help in estimating the amount of data being sent to the Cloud. In addition, the number of writers (e.g., sonographers and Reviewing Physicians) who will access the application to review and annotate these examinations to prepare patient reports need to be estimated. Lastly, the number of application users (e.g., Referring Physicians, clinical researchers, and other healthcare team members) needs to be estimated. A sample worksheet is shown in Fig. 6.7. All these need to be converted into data input, data stored, and data output from the Cloud on per year basis.

These estimates are rolled up in a separate table, as shown in Fig. 6.8a to see how data I/O, storage, and VMs (virtual machine instances) may grow over time. Given the current charges by the Cloud operators, these can be converted to expense figures as shown in Fig. 6.8b.

The Cloud cost estimates can be used to compare offerings from different service providers, such as Amazon Web Services, Google Cloud Platforms, Microsoft Azure, or any other provider to make a final selection. Currently, it is in vogue

	A	B	C	D	E	F	G	H	I	J	K	L	M
1		Q1	Q2	Q3	Q4	Q1	Q2	Q3	Q4	Q1	Q2	Q3	Q4
2	Year	2023	2023	2023	2023	2024	2024	2024	2024	2025	2025	2025	2025
3	Total Installed Units												
4	Patients/year (30/unit/month)												
5	File Size(GB)/patient												
6	Year	2023	2023	2023	2023	2024	2024	2024	2024	2025	2025	2025	2025
7	Writers/exam												
8	Data Input/qtr (GB)												
9	Readers/exam												
10	Data Output (GB)/qtr												
11	Total I/O (GB)/qtr												
12	Units/VMs												
13													
14													
15	Year	2023	2023	2023	2023	2024	2024	2024	2024	2025	2025	2025	2025
16	Storage($/GB/qtr)												
17	ml.m5.8xlarge												
18	BW (GB/qtr)												
19	Total$$s/qtr												

Fig. 6.7 A sample worksheet to estimate data input, output, and storage required in the Cloud

a

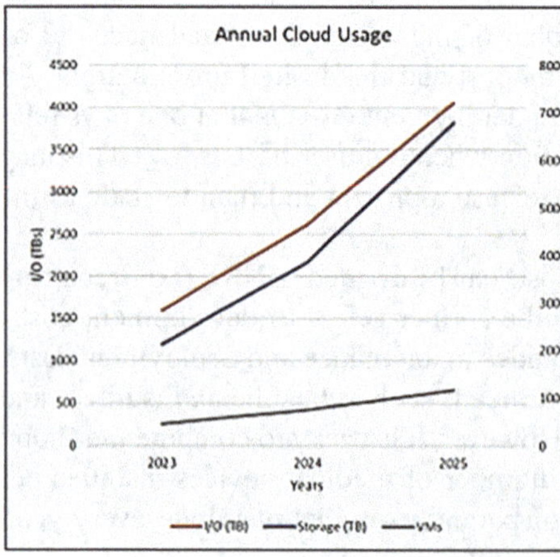

b

	Year1	Year2	Year3
	2023	2024	2025
Storage	$ 50,003	$ 95,162	$ 165,062
I/O	$ 271,728	$ 445,046	$ 688,867
VMs	$ 859,140	$ 1,407,132	$ 2,936,803
Total	$ 1,180,871	$ 1,947,340	$ 3,790,732

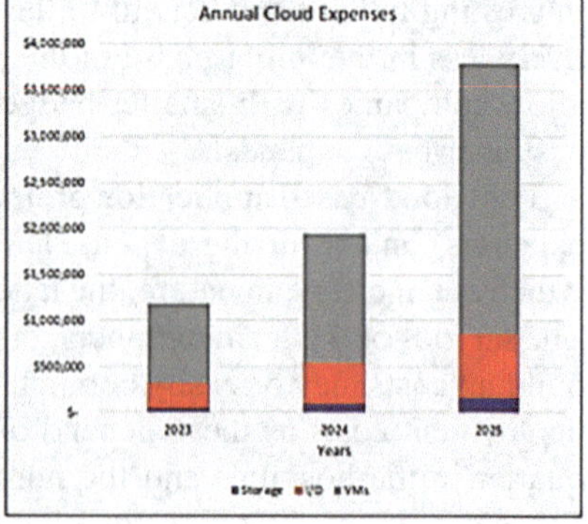

Fig. 6.8 (**a**) Data and compute growth in the Cloud and (**b**) Cloud expenses over time

currently to deploy multi-platform Cloud applications [3]. The goal here is not to necessarily lower the cost but to enhance the reliability of operations. The cost estimates are a basis used to determine the price of the subscription model for the application. Another input is the desired gross margin resulting in an overall profit for the enterprise. This will help to convince the management that this is a worthwhile project.

Let us now consider an example from a non-medical domain. Suppose there is a movie serving Website, such as Netflix, which could experience growth in a particular region. This entails locating the servers in the same or nearby region. This will ensure minimal latency for the new and existing customers. Also, it is important to identify potential bottlenecks, as the new customers may demand more of the existing or new movies. Serving more customers with the same set of movies will need duplication of network resources. Adding more movies to the repository would require additional compute servers to encode new movies, and storage devices to host them, while maintaining the older movies on the existing servers. Thus, it is important for a business to understand its users, their needs, and growth patterns. Furthermore, using spreadsheets various scenarios need to be worked out on what to host where. This ensures a better estimate of the cost, and it may be even budgeted in advance. Let us now work with numbers to understand how estimates may be prepared.

Suppose an application needs four EC2 (Elastic Cloud Compute) instances and S3 (Simple Storage Service), where the application running on EC2 requires 8 GB setup minimum. If the cost of each server instance is $0.10/h, and storage cost is $1.00/GB/month, then let us calculate the yearly budget for this organization.

Server instance cost/year = 4*$0.10*24*365 = $3504/year
Storage cost/year = 8*$1*12 = $96/year
Total outlay = Instance Cost + Storage Cost = $3600/year

Let us consider yet another business of online newspapers, magazines, or on-demand video services, where many users access read-only data through a Web application. If this business expects to grow by 50% per year, then its storage cost will not rise rapidly, and be taken to be a constant due to the read-only nature of operations. However, the server cost/year will grow by 50% to support 50% more users, bringing the new total to $5352/year. Now let us say another set of 50% users demand new magazines, which need new computers and storage, for the same proportion, requiring 50% more instances and storage. This will require new expenses estimated to be $8082/year, which is more than 2× of the first year's expense.

More users generally ought to be good news only if accompanied by a commensurate growth in the compute and storage infrastructure. Otherwise, the result will be similar to doubling the number of cars on existing roads, causing traffic jams slowing down all the users.

6.7 Optimizing Across Clouds

Since multiple Cloud service providers (CSPs) exist, it makes sense to be able to compare their offerings on a common set of metrics to choose the one that best meets a user's criteria. Furthermore, it may be desirable to have multiple accounts and make an application portable across CSPs to avoid lock-in possibilities. This is easier said than done, as each Cloud provider offers different sets of tools and different flavors of solutions, with an aim to attract and lock in their customers. Currently, there are three large public Cloud providers in the USA, as described in Table 6.1 below.

The above comparison is primarily SaaS based—the dominant platform for Cloud-based applications. Cloud providers' costs are changing ever so often, e.g., AWS has changed its prices 50 times since it was launched in 2006. Although other benefits are associated with the Cloud infrastructure, the low upfront costs and inexpensive payments for servers attract a large segment of customers and are always mentioned as a major incentive for Cloud adoption. It is important to understand the provider's pricing policy to make an informed decision on IT spending. To optimize the Cloud infrastructure's total cost, businesses need to explore the following options:

1. Size of VM (e.g., large, medium, or small in terms of memory allocation).
2. Number of VMs.
3. Likely duration of Cloud contract.

Table 6.1 A brief description of three large Public Cloud Service Providers in the USA

Provider	Offering name	Description
Amazon	AWS	A comprehensive, evolving Cloud Computing platform provided by Amazon.com. Web services are sometimes called Cloud services or remote computing services. The first AWS offering was launched in 2006 to provide online services for Websites and client-side applications. It enabled SaaS for the then emerging new ventures such as Netflix and Salesforce.com, etc.
Microsoft	Azure	Microsoft Azure began as a Windows server-based Cloud platform but has since grown to become a very flexible Cloud platform that enables developers to use nearly all popular programming languages and frameworks languages, or tools to build, deploy, and manage applications. According to Microsoft, Azure features and services are exposed using open REST protocols. The Azure client libraries, which are available for multiple programming languages, are released under an open-source license and hosted on GitHub
Google Cloud		A Cloud Computing platform by Google that offers hosting on the same supporting infrastructure that Google uses internally for end-user products like Google Search and YouTube. Google Cloud Platform (GCP) provides developer products to build a range of programs from simple Websites to complex applications

Specialized Cloud type	Infrastructure	Application platform	Application	Vertical-specific Cloud
Services provided, e.g., AI as a Service	Storage Computer Networking Management	Middleware Design Development Deployment Testing Operations	Vertical Applications Healthcare Retail Finance Manufacturing Enterprise Consumer	Healthcare Retail Finance Manufacturing Others

Fig. 6.9 Evolution of Cloud service providers

4. Require resources, such as compute, storage, networks, configuration and security framework, etc.
5. Pricing models.

Pricing models vary across AWS and Azure and other providers. The factors mentioned above provide a starting point to compare offerings from different CSPs. In the coming years, focus may shift from Cloud Computing to specific Cloud services. As an example, MS office on Windows 11 now uses a monthly or yearly subscription model. The specific services may be for the domains such as logistics, supply chain management, social media, media processing, enterprise management services, HPC, machine learning and AI applications, gaming, etc. Specialized Cloud will begin to emerge providing services like those highlighted in Fig. 6.9.

Changing Cloud Service Providers (CSP) is a nontrivial task. It is not as simple as changing a cellphone or Internet service provider. The main reason is that a Cloud App is often tied to a particular configuration for a class of VM, OS, APIs, data storage, etc. Cloud migration needs careful planning and execution considering the following three aspects:

1. Virtual or abstract machine level with Operating System.
2. Use of a CSP-specific APIs and services.
3. Data migration.

To avoid any surprises later, many App developers start with a multi-Cloud deployment strategy [3]. This helps in migrating among the CSPs as and when it becomes necessary.

6.8 Cloud Service Level Agreements

A service-level agreement (SLA) is a contract-negotiated and agreement between a customer and a service provider. The service provider is required to support incoming requests from a customer within the negotiated quality of service requirements for a given price. A server in Cloud needs to operate with variable load dynamically provisioning computing resources to meet the SLA. Note that optimal resource utilization is a nontrivial problem to tackle. For example, offering guaranteed network bandwidth or data read-write speeds for an application is a key optimization problem of significant difficulty. These tasks are generally conducted on shared resources, such as a network interface card or a memory controller on a server with multiple virtual machines or applications accessing data over the same set of physical wires. This may result in a conflict of demand for resources and their availability of a shared multi-tenanted server resulting in unexpected wait times. The only way to guarantee a certain bandwidth is to reserve it in advance. Note that this could mean leaving parts of it unused till an application request for it. However, that leads to unused capacity and higher costs for the Cloud operator. Hence, they often over-allocate the same resource to multiple customers and avoid specific SLAs (like the airlines marginally overbook their flights) with precise guarantee of quantified performance.

Most Cloud SLAs specify only uptimes [8, 9], stating that their compute servers, storage devices, and networks will be available for a certain percentage of the time. Service is measured by “9 s” of availability, such that 90% has one 9, 99% has two 9 s while 99.9% has three 9 s, etc. A good high-availability (HA) package with sub-standard hardware has three 9 s, whereas an enterprise class of hardware with a stable Linux kernel offers five or more 9 s. Below is an example of typical availability levels in Table 6.2.

At this stage, let us review some relevant definitions in this context:

“Downtime period” means a period of 60 consecutive seconds of downtime. Intermittent downtime for a period of less than 60 consecutive seconds will not be counted toward any downtime periods.

“Monthly Uptime Percentage” means total number of minutes in a month, minus the number of minutes of downtime suffered from all downtime periods in a month, divided by the total number of minutes in a month.

“Maximum financial credit” refers to the aggregate number of financial credits issued by a Cloud service provider to the customer for all the downtime periods in a single billing month. For instance, Google computes it such that it will not exceed 50% of the amount due from customers for the covered service for the applicable month. Financial credit will be in the form of a monetary credit applied to future use of the covered service and will be applied within 60 days after the financial credit was requested.

Microsoft Azure SLA offers commitments for uptime and compensates for downtime by offering service credits. An example of virtual machine connectivity,

Table 6.2 Examples of availability by the 9 s

9s	Availability	Downtime/year	Examples
1	90.0%	36 days, 12 h	Personal computers
2	99.0%	87 h, 36 min	Entry-level business
3	99.9%	8 h, 45.6 min	ISPs, mainstream business
4	99.99%	52 min, 33.6 s	Datacenters
5	99.999%	5 min, 15.4 s	Banking, medical
6	99.9999%	31.5 s	Military defense

Table 6.3 Example of MS Azure Credits in event of downtimes

Monthly uptime percentage	Service credits
<99.95%	10%
<99%	25%
<95%	100%

Table 6.4 Example of Google Cloud Credits in event of downtimes

Monthly uptime percentage	Percentage of monthly credits
99% to <99.95%	10%
95% to <99%	25%
<95%	50%

which encompasses server and network uptimes, is given below, followed by credits offered for downtime in Table 6.3:

For all the virtual machines that have two or more instances deployed in the same availability set, the connectivity to at least one instance is guaranteed 99.95% of the time. For any single instance, virtual machine using premium storage for all operating system disks and data disks, Azure guarantees virtual machine connectivity of at least 99.9%.

Similarly, Google Cloud Platform offers a stringent SLA and credits back for any downtime as shown in Table 6.4:

Besides the availability, SLAs should specify downtime, mean time to repair, and mean time to respond. Scheduled maintenance of updating systems doesn't generally count as downtime.

Some Cloud service providers offer multilevel SLAs, e.g., Rackspace offers two level options (managed or intensive) for support [10] as described below:

- 100% network uptime guarantee.
- Within 1 h hardware replacement.
- URL, port availability, and hardware monitoring.
- Managed firewall and VPN access.
- OS patching.
- Bandwidth and backup performance utilization.

The following suite of services is offered only to intensive-level customers at a higher cost by Rackspace:

- Server virus scanning.
- Guaranteed response times for support requests.
- Advanced system performance and device monitoring.
- Custom configurations.
- Performance trends.

In addition, the following optional services are available for an additional cost:

- DDOS (Distributed Denial of Service) attack mitigation.
- Managed security services.
- Encrypted backups.

However, Rackspace customers may not get predicted, measurable, and quantified performance levels. Similarly, Amazon offers dedicated servers, but customers must be content with whatever performance level becomes available depending on the workload and resource availability constraints. This is one of the key limitations currently observed in the Public Cloud. Data security and application performance are at best considered to be a shared responsibility between the developers, users, and the Cloud Service Providers. A detailed discussion is beyond the scope of this book but can be found here [3].

6.9 Cloud Project Roles and Responsibilities

In the previous chapter, we listed different people needed on a project team and their respective roles. This needs to be revised in the context of Agile [2]-based project since it is ideally suited methodology for developing a Cloud-based software product.

Below we list the primary and key roles for executing Agile processes:

1. The Scrum Master.
2. The Product Owner.
3. DevOps Engineer.
4. The Development Team.
5. Validation Team.

The last role relates to group behavior and operations. Essentially, one expects that the team folks subscribe to the Agile values. In short, they are willing to follow the Agile process and keep focus on the deliverables. The dynamics of the group is, however, driven by the Scrum master and the product owner. So, we shall discuss these two roles in more detail.

6.9.1 Role of the Scrum Master

A Scrum master is a mentor, coach, and a manager all rolled into one. A mentor who is very knowledgeable and can, therefore, be looked up to by all, i.e., the team, as well as, the product owner and above all by sponsors. She is trusted and respected by the sponsors for her abilities and managerial skills. Invariably, such a person would have been product owner a few times in the past. She is mature enough to understand group dynamics and, therefore, can act as a coach as well. She ensures that the team continues to believe in the Agile values. To reinforce this, she periodically underlines the collaborative spirit in all the meetings in a subtle way. She is aware of the best practices in management and encourages collaboration within the team without sacrificing the top management priorities so that the focus on the deliverables remains intact.

One key responsibility she shares with the product owner is translation of sponsor's vision of the product. If she synchs up well with the product owner, her task would primarily be a mentor.

6.9.2 Role of the Product Owner

A product owner is basically a technical manager who is a very knowledgeable person with a strong technical background. He should be able to translate and understand the sponsor's vision of the product. He should be able to assess and estimate time and effort required for the tasks in the project. He is expected to evolve strategies, allocate resources and ensure time boxed deliverables. As for the management—his biggest challenge is to manage product backlog. If he slips on product backlog—he has to be able to devise a strategy to catch up on sprint and also manage velocity. He directly reports to the Scrum master.

The roles discussed above are the two key individual roles that matter in Agile process management. It would be beyond the scope to discuss support tools for Scrum master and product owner. There are many SW tools to help execute and drive Agile processes [2].

6.9.3 Role of the DevOps Team

Just as the Agile methodology emphasizes a degree of communication between stakeholders, the DevOps encourages the development and deployment teams to work collaboratively. Note that the development team may be located in a different location or geography (such as India) while the application's deployment may be happening in organizations in a different geography—for example, hospitals located in USA. They typically allow data access only within the national boundaries due to the laws such as HIPPA [6]. Then it falls on DevOps to bridge this gap as nobody from the development team can login to a hospital site in the Cloud. DevOps resolves system engineering and integration issues for the end users and customers.

It has also been found that the DevOps offer benefits in terms of better version control, configuration management with more stable releases and predictability [3]. The DevOps team use tools to generate a slew of microservice container options in the Cloud for deployment. In the Cloud context, DevOps help rolling out new releases with evolving and emerging specifications of the network and server configurations.

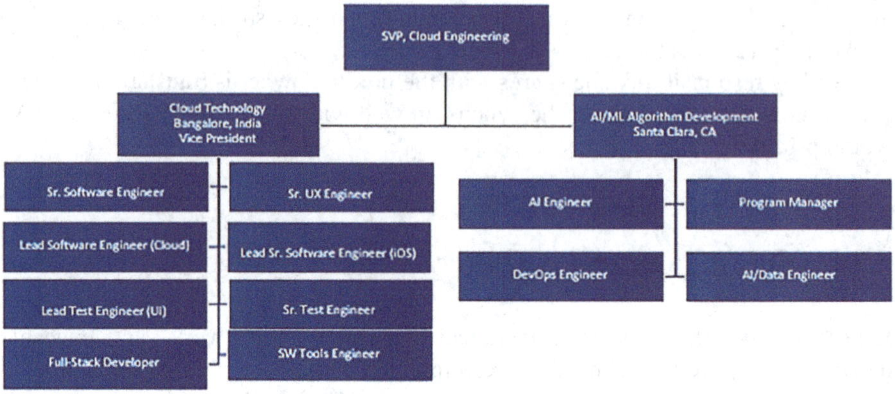

Fig. 6.10 A representative Organization for Cloud App development and testing team

6.9.4 Overall Project Setup

For the medical App discussed earlier, it was imperative that DevOps role was located in the USA. It allowed the team to serve the most urgent customer needs in the same time zone and follow the HIPPA guidelines. However, rest of the software development and verification teams were based in India. Since the medical device hardware and sales teams were also based in the USA, we needed a local program manager (PM) to coordinate customer demos, and to work with the PMs of hardware team. In addition, since the hospital data was collected and kept in the USA, a fair amount of analytics was done on it for which we needed a couple of engineers closer to the customer locations. In Fig. 6.10, we have a representative breakdown of operations across multiple geographies.

Figure 6.10 should be treated as a representative organizational structure. Depending upon the nature of the project this may be tweaked—yet we believe it could serve as a generic template.

6.10 IoT Project Roles and Responsibilities

IoT is a classic paradigm where both DevOps and Agility are required. Since multiple domains may be involved in an IoT-based project, Scrum plays a crucial role. DevOps also plays a critical role in amalgamation of technologies from different spheres. Therefore, synchronization of subgoals becomes very important, requiring both Scrum and DevOps in an IoT-based Cloud project, as follows:

1. Scrum Master ensures that Scrum events and processes run smoothly. These include Daily Stand-ups, Sprint Planning, Reviews, and Retrospectives, creating an environment where domain-based IoT teams collaborate effectively.

Additionally, the Scrum Master ensures compliance with the Cloud, such as scalable APIs.
2. DevOps Engineer fosters a culture of collaboration, breaking down traditional barriers between development and operations. When applied to IoT, this collaboration ensures that device development aligns with Cloud's operational requirements, leading to more reliable and effective solutions.

An IoT project often requires the following additional participants, with their roles and responsibilities listed below:

3. IoT Architect conducts research, designs, and implements prototypes for cost-effective IoT-based solutions. These include Internet connected devices engaged in exchanging data.
4. IoT Engineers develop sensors, devices, and software to build an IoT-based product. They also design, code, and test features of IoT devices and provide solutions to network and platform connection issues.
5. IoT Project Manager facilitates real-time collaboration and communication among project stakeholders.
6. IoT Operations Manager keeps track of all the devices on the network, ensures these are functioning properly, provides updates and security patches.

6.11 Project Monitoring and Analytics

A project may start with a plan and a certain budget allocation, but it is important to track its progress during execution and deployment with various metrics for ensuring success. During the implementation stage, the metric may be as simple as list of features that have been completed and tested, list of bugs found during testing and fixed, etc. During deployment, it may collect data such as the number of customers, their usage patterns, data uploads, accesses, etc.

All these metrics need to be collected periodically—say weekly and shared with management to gain confidence and receive continuous support for the ongoing project. Many projects face cost overruns due to new requirements and unexpected technical or non-technical challenges. It is important to communicate serious concerns early enough to the management so that timely corrective actions can be taken. Sometimes, management is faced with a delay and natural response may be to add more resources, which may not be necessarily the right thing to do. It is well known that adding new team members to a delayed (or derailed) project delay (or derailed) it further [11]. If a project is late, then the customers may choose a competitor's product instead. Any surprises during execution should be discussed with the data on the causes, options with pros and cons to decide on what actions can be suitably taken to keep the project within the desirable and feasible boundaries.

After a project is finished, it is important to take some time to conduct a study to learn what went right and could have been done better to avoid making the same mistakes again or in the next project. A team builds relationships and gains tribal knowledge during every project's execution. This is important to carry over to the next project so the team's performance can improve over time.

Another aspect of data collection and metrics happens after the Cloud App is completed and deployed across multiple hospitals. Each hospital will have its own secure portal in the Cloud that will contain PHI (Private Health Information) identifying the patients to the physicians to offer proper treatment. However, there is an additional opportunity to share the knowledge of these treatments among the medical community provided a data lake is formed [12]. This would require data de-identification to comply with HIPPA protocol. Once PHI is removed then patient examinations, physicians' diagnoses, and treatments can be shared across hospital boundaries. Removing PHI is necessary but not sufficient. It is important to group together multiple examinations of the same patient to observe trends of disease progression and effects of treatment or side effects for comorbidity over time. This process of de-identification is implemented using tokenization, where PHI is replaced by a hashed token, so the same patient's information will generate the same token masking the patient's identity. Note that tokenization protects a patient's privacy, as reverse engineering from a token to PHI data is intractable.

IoT operations involve a network of connected devices that enable real-time monitoring and data analysis [13]. This offers project managers additional visibility to make informed decisions, streamline processes, and enhance efficiency. IoT also enables managers to monitor the performance of the equipment used in the project. For example, if a server or network's load suddenly increases, project managers can immediately investigate the issue and take appropriate action to prevent outages or downtime.

By integrating IoT devices with project management software (such as Jira [14] or MS Project [15]), project managers can receive updates on task completion, identify bottlenecks, and allocate resources effectively. This visibility enables better coordination among team members and ensures that project stays on track.

Once all the project management data is collected in a large repository, data mining can be done. In general, data mining models have specific objectives. The data mining exercise may take one or more of the following forms:

1. **Predictive:** to make a prediction about values of data using known results found from different datasets.
2. **Descriptive:** to identify patterns or relationships in data. Unlike the predictive model, a descriptive model serves to explore the properties of the data being examined. Generally Clustering, Summarization, Association rules, and Sequence discoveries are used to unveil hidden patterns in the given datasets.

Using the above techniques, all the existing examination datasets covering a wide spectrum of patients can be grouped for use in future projects for classification tasks. The patient data may be marked as healthy or unhealthy patients. In addition, the analytics can help in enlarging the range of treatment regimens for unhealthy groups, which can be divided into different diagnostics categories. Their common symptoms can be extracted, based on parameters such as heart rate, blood pressure, and sugar levels. Then a newly submitted examination can be quickly measured against the parameters for each known group to ascertain if the new patient is healthy or what is the likely nature of attention needed. This can be presented to physicians as a suggestion with some specific examples.

6.12 Summary

In this chapter, we explored Cloud-based product planning and strategies for its execution. A product's life cycle begins with a clear understanding of different user personas, their requirements, pain points, and expectations. The next important steps are the selection of a Cloud Service Provider (CSP), resources, and cost estimations based on expected usage patterns. The chapter elaborates some of the methods used for Cloud-based product monitoring, management, and use of data analytics.

Points to Ponder

1. Why is it important to enumerate different users' personas for an application?
2. How does UML help with TDD and BDD tests development?
3. Why is it important to build futuristic cost models for a Cloud-based Application?
4. What are the customer and business benefits of data lake?
5. How does IoT-based project require an extension of Cloud team roles?

Answers

1. Why is it important to enumerate different users' personas for an application?
   (a) Once we have enumerated different types of users and stakeholders, we need to proceed to design a product to meet their requirements. In particular, their pain points need to be understood to make the product truly user friendly.
   (b) Often users are not able to articulate their needs precisely, so these have to be teased out by asking pertinent questions or by observing their workflows in action.
   (c) Once all user needs are documented, these have to be ranked in the order of importance from both businesses and users' point of view. These may be documented in an MRD (Marketing Requirements Document).
   (d) A PRD (Product Requirements Document) generated from the MRD, by considering available investment, resources, and execution time.
   (e) Also, note that the system architecture is proposed from possible alternatives to realize the finally selected user requirements for a product.
2. How does UML help with TDD and BDD tests development?
   (a) UML captures structural and behavioral representations for a system being designed.
   (b) A structural diagram shows the modules and their interfaces. Each module needs to be tested at unit level as well as an integrated product to ensure that the given requirements can be met.

(c) A behavioral diagram shows the data and control flow, which drives use case-based testing.
(d) The two together nearly check out all the user requirements.

3. Why is it important to build futuristic cost models for a Cloud-based Application?

(a) A business application is written to meet the users' needs and drive profits for the investors.
(b) Unlike hardware or enterprise software, a Cloud-based application may be deployed on any third party infrastructure. Its variable costs include storage, networking, and server usage.
(c) These future costs and trends need to be projected to ensure that the App will continue to generate profit for the business. Otherwise, it will not be sustainable business.

4. What are the customer and business benefits of data lake?

(a) The customers in a Cloud App can only access their own data, which may be limited in scope and variants. One can make better decisions by using other customers' anonymized data.
(b) A Cloud App provider can aggregate data from multiple customers to assess their behavior and make better business decisions. It is easy to release to several example scenarios from domains like airlines ticket pricing, hospitality business, and to auto makers who build up their car inventories with the right models and colors, etc.

5. How does IoT-based project require an extension of Cloud team roles?

(a) IoT-based project requires DevOps and Agility. Since multiple domains are involved, Scrum plays a critical role.
(b) DevOps Engineer fosters a culture of collaboration, breaking down traditional barriers between development and operations.
(c) Scrum Master ensures that Scrum events and processes run smoothly.

References

1. Covey, S. *The 7 habits of highly effective people: 30th anniversary edition*. Simon and Schuster. ASIN: B07WF972WK.
2. Bhatt, P. C. P. (2021). *Software design, architecture and engineering: Concepts and practice*. Prentice Hall India, Easter Economy Edition. ISBN-978-81-951611-4-0.
3. Sehgal, N. K., Bhatt, P. C. P., & Acken, J. M. (2020). *Cloud computing with security and scalability*. Springer. https://link.springer.com/book/10.1007/978-3-031-07242-0
4. https://techbeacon.com/enterprise-it/5-steps-building-Cloud-ready-application-architecture
5. https://aws.amazon.com/Cloudwatch/
6. https://www.hhs.gov/hipaa/index.html
7. https://Cloud.google.com/storage/docs/encryption
8. https://azure.microsoft.com/en-us/support/legal/sla/
9. https://Cloud.google.com/pubsub/sla

10. https://www.rackspace.com/en-us/managed-hosting/service-levels
11. Brooks, F. (1979). *The mythical man month*. Addison Wesley.
12. https://en.wikipedia.org/wiki/Data_lake
13. https://teamhub.com/blog/real-time-iot-project-tracking-for-effective-project-management/
14. https://www.atlassian.com/software/jira
15. https://www.microsoft.com/en-us/microsoft-365/project/project-management-software

Chapter 7
Communications: The Life Blood of a Project

7.1 Introduction

Once the need for a new Cloud-based application is established, a project team is put in place. The initial task for the team is to create an execution plan and initiate its implementation. It is important to track a project's progress and keep the management duly informed, so that the funding and support continues. Often something or the other changes with the passage of time, and unforeseen issues or difficulties may be encountered which can delay the project or require additional investment. No project is an island as the intended application will need to function in an ecosystem of other hardware and software components. This may require creation of interfaces to the existing products to provide data and control information. This is the background in which the communications assume importance in implementation and delivery of a successful project.

We shall explain this aspect of communication in the context of NovaSignal's medical App [1]. We begin this chapter by first exploring device-based communication, and later discuss the importance of communication between humans, which is essential for a project's success.

7.2 Device-Based Communications

It is common knowledge that medical practitioners use MRI and CT scans to study a patient's brain. The high energy waves are used for scans because nature posited the human brain inside a pretty thick protective bony skull. However, it is also known that there are small openings in human skulls. The App described here uses these openings to direct ultrasound signals in 2 MHz range to measure the velocity and extent of blood flow in the blood vessels. The reflections of these waves are

P. C. P. Bhatt et al., *Project Management in Cloud and IoT Applications*,
https://doi.org/10.1007/978-3-031-99427-2_7

measured using the Doppler formula (recall your high school physics days). The doctors interpret these ultrasound measurements to assess if the patient is suffering from some form of brain disease manifested in the blood flow. This can help them to determine what ought to be the line of treatment.

Let us explain the above in the context of one Cloud-based application developed by NovaSignal [1]. The App takes data from an ultrasound machine as shown on the bottom left side of Fig. 7.1. Prior to the introduction of a Cloud Application, data from the NovaGuide ultrasound machine was transferred to PACS (Picture Archiving and Communications System), which can typically store only static images (as JPEG files), but no sound or video data stream. Sound signals are important in any ultrasound examination. An expert physician or sonographer can detect the sound of blood flow to detect abnormalities in a patient. This is because laminar blood flow generates smooth repetitive sound bites with the heart beats. Whereas, with clots (aka emboli) or air bubbles produce a distinct glitch or non-laminar noise in the blood flow sounds in the brain or elsewhere in the human body. Thus, it is critical to save the recordings of ultrasound for future reference and treatment of a patient at a location where the authorized users can access it anytime anywhere. It is imperative that the new interfaces are well defined to ensure that the Cloud application can get the necessary input feeds. The data from NovaGuide ultrasound machine is encrypted and sent over the Internet using HTTPS/TLS 1.3 protocol to a secure Cloud portal, where it is stored using AES-256 encryption algorithms. Once the data has been uploaded then all the authorized medical practitioners can access it anytime, anywhere using this mobile App.

The dataset collected from each patient's examination can be a few Gigabytes or larger in size, depending on the duration of the ultrasound test. The information is divided into two parts, containing a patient's Private Health Information (PHI) and

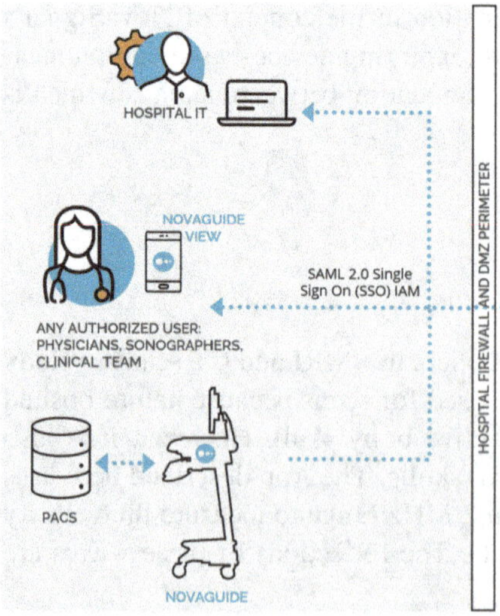

Fig. 7.1 Data and Control flow from a medical device to the Cloud [1]

the medical data itself. Both need to be handled while strictly complying with HIPPA. Also, a unique portfolio is created in the Cloud for each patient, so that all the future examinations for that patient can be linked together.

Whereas physicians search the database using PHI, a medical researcher needs to access only de-identified examination data, i.e., devoid of PHI. This means that the application should be able to determine different access rights based on the type of user. A hospital's IT manager, therefore, needs to set up different kinds of accounts and monitor their usage patterns. Every data access needs to be logged, and logs must be saved for at least 7 years in compliance with HIPPA. Note this is pertinent only for the data access logs. It may be desirable to save medical data for the lifetime of a patient.

As we shall later explain in this chapter, there are two modes of Cloud Application: a batch mode to upload medical examination data for reviewing it later, and a real-time mode to get alerts while an examination is in progress. The latter is useful in mission critical situations while a patient is being driven to a hospital in an ambulance, so that the medical team can prepare to take care of the patient on arrival.

7.3 Limits of Technology

While it is desirable to save all aspects of a patient's examination, current PACS systems store only JPEG images and PDF reports. This limitation can be readily overcome in the proposed Cloud-based application using elastic storage, for any-time anywhere access over the Internet by medical personnel. However, it also places an additional burden of complying with privacy and security regulations. For instance, HIPPA or GDPR restrict access to medical data and require that all logs should be kept for up to 7 years.

Such requirements need to be clearly stated during the project requirements gathering phase as described in previous chapters. Furthermore, the real-time nature of a medical examination requires immediate communications for critical patients. This may include checking a patient's condition during transportation to a hospital, communicating it to emergency room doctors who can even decide a suitable location for treating the patient enroute. This scenario is depicted in Fig. 7.2.

For example, in these cases it is important to differentiate between brain strokes and brain hemorrhage. Both manifest very similar medical symptoms. The former represents a blockage of blood flow due to an embolus, which starves the brain tissue of vital oxygen. The latter represents leakage of blood flow due to an aneurysm, which also starves the brain tissue of vital oxygen. The first case is often treated with anti-coagulant such as aspirin, or blood thinners to restore the blood flow. However, if administered wrongly to a hemorrhagic patient, it will cause additional leakage of blood sometimes leading to death. Thus, it is important to diagnose the patient correctly on the way to the hospital. This requires an ultrasound machine hardware to be present in the ambulance in live communication with the medical care team to interpret the results of an examination. If the data from any medical

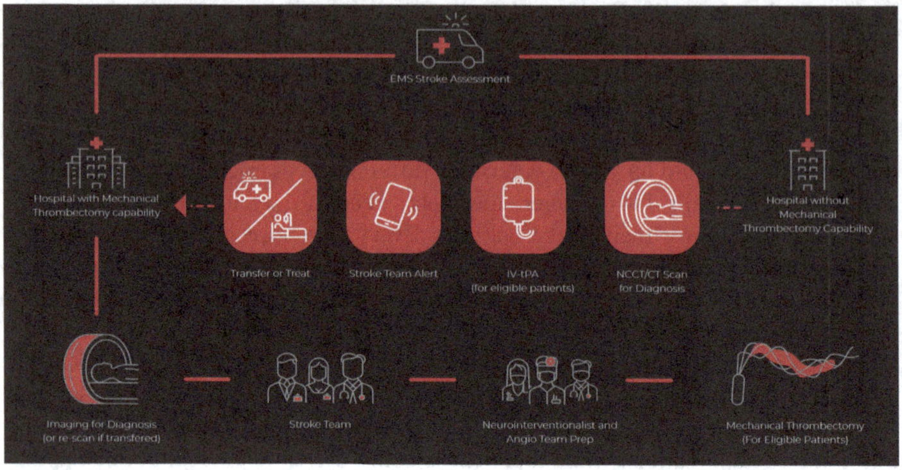

Fig. 7.2 Clinical Decision Support System for brain diseases [2]

device is uploaded to a hospital's Cloud portal, then notifications are sent to the medical care team. In that case, the latency of both the segments needs to be considered. Latency in both cases can be minimized by choosing a datacenter in a geographical location closer to the hospital.

We will not delve into the physics of doing an ultrasound in a moving vehicle, despite the vehicular motion and vibrations due to the road conditions, as it is beyond the scope of this book. However, for the hospital personnel to decide the course of treatment it is important to gather this data and reliably communicate it to Cloud. Thus, low latency, reliable, and secure data transmission requirements are built into the planning stage of the Cloud Application and its deployment decision.

The team which is implementing this project was in a different geography. This is where a project manager and DevOps engineers play an important coordination role. During clinical trials, field data was collected to be shared with the project team. If this resulted in some requirements modifications, then these needed to be accounted for in the Agile-based execution of the project.

Let us now explore some aspects of human communications needed for the success of any Cloud application project.

7.4 Maintaining Team Morale

It is said that 80% of communication is non-verbal, which includes written, and information picked from face-to-face interactions. Since a remote execution precludes the latter, it may be pertinent to record videos of field trials, interviews with the users to fine tune the user interfaces of a Cloud-based application. This is because if the users find a feature clueless or hard to use, then they will tend not to use it regardless of the goodness of the underlying algorithms or value of the results. This may result in some disappointment among the development team members, who may have worked hard and long hours to deliver a feature that users prefer to let go by for the lack of usability.

In such an environment, it was important to maintain the morale of the team in a remote geography. This effort started with daily scrum calls that may be at inconvenient hours for the other geography. In the case described here, the development and testing teams were in Bangalore, India, while the deployment and project managers were in the Silicon Valley, USA region. Given this background, a good time to hold these daily scrums was mornings for the India teams. Such an arrangement shows good faith in the effort being put in by the development and test engineers. It also showed USA based managers in good light as they are willing to sacrifice their evening or night times to accommodate time zone differences. The morning scrum, in fact, helps in setting the agenda for the day at the Indian end.

Another step that helped to maintain team morale was frequent visits by high-profile managers from the headquarters, as it showed commitment and value being placed on the remote team's work.

Lastly, any customer feedback, positive or negative, was shared on a regular basis so that the course correction were made resulting in the successful agile release of the project.

7.5 Tracking Project Progress

As one famous saying goes: "Rome was not built in a day." Any worthwhile project will take many months or even years to complete. It is important to break down the problem to tackle it in stages. For this Cloud-based application three generations of product releases were planned right at the onset. The operations of the three stages are described next.

1. **Remote access to data**: In this stage, the application can ingest data from an ultrasound machine located in a hospital to the Cloud. Then using a Web-based client application, physicians, and care giving team members should be able to access this data as desired.
2. **Real-time examinations**: At this stage, the product should be able to get the data as a real-time stream emanating from a patient's examination. This was critical for post-operative patients. For example, a heart operation may be successful, but during the healing stage, clots may manifest in response to the surgical wounds. If one of these clots breaks off and travels in the blood stream, it may result in a stroke for the patient, even though the heart surgery itself may well be deemed successful.
3. **AI/ML-based interpretations**: This was for both a real-time and uploaded examinations. The former requires continuous monitoring of a patient. Since physicians cannot stand for hours at the bedside waiting for a clot to appear.

Application requires auto-detection of an embolus and sending a real-time alert to the care giving team. The concerned doctors can get a notification on their phones and offer immediate treatment options. For the uploaded examination, an automatic classification system can mark the patient as healthy or otherwise. The second category can have detailed sub-classifications such as stroke, hemorrhagic, or otherwise abnormal blood flow. In this case, the system should be able to point out specific periods (in a long examination report) where any abnormality was observed. The goal of this system was to direct physicians' attention to the problematic areas and speed up the diagnosis process. Note the machine learning model can be trained on a previously labeled dataset. Such a system improves incrementally through a learning process over time.

Once the above three product generations were identified and agreed upon, the implementation work began with a development and deployment pipeline in place. Note that while a product was being moved into production, work on the next generation could start.

During a project's execution, it is imperative to track its progress. If requirements change based on a field trial or new user inputs, then these need to be incorporated in the requirements document. When a request is received, the people who own the requirement and its scope need to be consulted. The change can be incorporated only through consensus. One of the key considerations in this context is to maintain a traceability matrix as shown in Fig. 7.3.

The X marks in Fig. 7.3 indicate the relationship of a certain module with the requirement in SRS statement. In fact, it defines a traceability matrix. In general, the module under reference in a traceability matrix may either be hardware or software

Rqmt. No and description	HW Mod-1	HW mod-k	SW mod-1	SW mod-m
1.1 Functional reqmt. – 1		X		
1.1.1 Introduction				X
1.1.2 inputs	X		X	
1.1.3 Processing				
1.1.4 Outputs		X		
……				
1.n Functional reqmt. - n		X		
1.1 External I/f reqmt.	X			
1.1.j External communication I/f			X	
…..				
p.1 Quality characteristics				X
….		X		X
r.1 Ranked for importance				

Fig. 7.3 A typical traceability matrix organization [3]

module. It shall be quite in order to identify team ownership from the traceability matrix. That would facilitate change management to be carried out in an orderly manner.

The final document itself may be sectionalized to adhere to an internal documentation standard. This is the final document that forms the basis for the SW architects and designers to create a conceptual framework for implementation. To support these activities, many frameworks and tools have been developed over time with various forms of representation. The requirements and specifications are often captured in textual and graphic forms. The basic idea is to convey the intent of the proposed Cloud-based software application project. Its progress can be tracked as shown in Fig. 7.4.

One textual mechanism is to use a story-based framework [3] to support Agile SW development. Graphics are useful in depicting communication between participating agents (behavioral description). It also captures time and event sequence of operations very well. The communication between devices or SW entities gives a better view of the system. Whenever there is a perceptual visualization of behavioral aspects of a system—then we have a model of that system. There are many SW tools and utilities that generate documents rich both in textual and graphical forms. These describe SRS development methodology along with some examples.

The progress of a project can be monitored with regular code reviews, tests passed, and features completed. The notion of extensive SW testing is in-built in Agile. It uses a metric called RTF (Realized-Tested-Features) on a continuing basis. The periodic evaluations using RTF are aimed at deriving customer satisfaction [3].

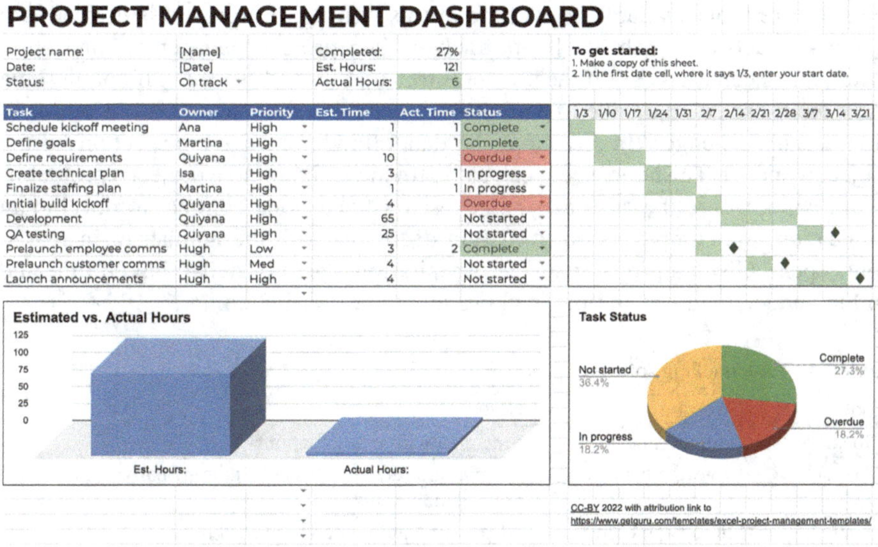

Fig. 7.4 A sample project tracking dashboard [8]

Progress can be mapped against the original plans and milestones on the roadmap. If needed, project delivery dates may need to be renegotiated. Generally, if a project is slipping, adding more resources to it is generally a bad idea as new people will need to be trained and assimilated. Adding more people to a delayed project likely will delay it further [2]. A better method is to delay the timeline or drop some non-critical features. This is the concept of MVP (Minimal Viable Product) and should always be kept in view to see which are the must-have-features.

7.6 Collecting Inputs for Data-Driven Project Management

Currently, many tools are available to capture and display data on project execution. However, of particular importance are the tools that track and provide analytics into hours spent by the team on the project. These tools help to monitor resource availability and utilization, and tools that track scope increases to project deliverables. Regardless of the tools used, applying learnings from data gathered during the project increases the probability that the project will ultimately end in success and allows for improved predictability of an organization's future projects. This process is referred to as data-Driven Project Management.

Tracking hours spent by the team members is valuable information for the project manager, whether a tool is used or not. When a project starts, and as it progresses and evolves, estimates for hours needed to complete tasks can be refined. Software tools can be found that make it very simple for a team member to "start" and "stop" tracking their time against their tasks, eliminating the need for that person to estimate how much time they spent after the fact. See Table 7.1 for a list of time tracking software tools available at the time of this writing. If a project is divided into sub-tasks, each of which has hours tracked, then project management and other stakeholders can easily see if the project is trending towards success. If not, in some cases the cause could be due to poor planning, and in other cases it could be due to unforeseen challenges. In any case, the data allows for course correction by the project manager. The hours spent can also provide insight into a project's profitability. In cases where the customer has chosen a fixed-price contract to develop an application, the company's profit margin decreases with the number of increased hours spent. Given a project manager's aspirational goal of delivering on time and within budget, data-driven time tracking should be considered.

Table 7.1 Time tracking software tools

Tool	Website	Notes
Clockify	clockify.me	Free plan with many features
Harvest	www.getharvest.com	Simple interface, integrates with third-party project management tools
Timely	timelyapp.com	AI powered automated time tracking
Toggl Track	toggl.com	Free for up to five users, easy to use
Tracking time	trackingtime.co	Built-in project management feature

Another common conflict that can affect efficient execution of a project is that of resource availability. Whether the resource is a member of the project team or a lab resource such as a thermal chamber for environmental testing, it is important that the entire organization understands resource constraints. Organizations should always understand the priority of various projects and the resources assigned to those projects. When a single resource is assigned to multiple projects over the same time, a clear plan should be used to determine the ultimate use of the resource. With accurate data input into resource tracking or project planning tools, reports can be generated that highlight conflicts. This knowledge can be used to pre-emptively avoid resource conflicts. Resource management and project planning software platforms often overlap in features and can be used for both. See Table 7.2 for some available resource planning software.

Projects are often subject to scope creep. This is when stakeholders, either internal or external, add or modify features to the project deliverables that were not previously mentioned in the project charter. Examples of internal scope creep include project engineers attempting to achieve higher performing specifications than the customer has requested given a desire to produce the best possible product or project managers agreeing to additional and time-consuming testing requests from a client to improve the client relationship. External scope creep typically comes in the form of customers changing or adding requirements in the middle of a project. The important point is that any scope creep must be identified, quantified, and tracked. At a minimum, any additional scope in a project should be written down and considered for additional analysis. This analysis could include a detailed estimate of resources needed and the impact on the cost and delivery date of the project. By investigating the impact, timing and acceptance or rejection ratio of scope creep, a project manager can more easily manage expectations for the project across all stakeholders.

Interestingly, IoT devices themselves can be considered for data collection, tracking, and automation in project management [4]. For example, IoT devices such as Lightbug's GPS tracker [5] can be used for asset monitoring and Aatmunn's Safety Control Center [6] can provide safety alerts for workers at construction sites. An organization should carefully weigh the pros of implementing IoT in project management against the cons. With real-time monitoring of project issues and deliverables, the project manager can react quickly to keep a project on track. Using devices to monitor productivity can lead managers to better decision-making when assigning team priorities. IoT devices can also help automate mundane or less

Table 7.2 Resource tracking tools

Tool	Website	Notes
Airtable	www.airtable.com	Many templates provided, opportunities for third-party integration
Asana	asana.com	Free basic plan, integrates with time trackers
Smartsheet	www.smartsheet.com	Provides project visualization and automation
Wrike	www.wrike.com	Offers AI recommendations, visual dashboards

technical tasks, allowing employees to focus on critical, high value work outputs. However, these benefits must be considered against the actual and perceived drawbacks when IoT devices are used as a project management tool. Data security and privacy are a matter of major concern given the amount of data, sometimes personally identifiable, that will be collected and transmitted over a network throughout a development project. Whether or not IoT devices are used as project management tools will ultimately depend on a risk analysis by the company keeping project team culture in mind.

7.7 Stakeholder Management

A product has many stakeholders besides its managers and implementation team members. These include the executive team of the company, its customers and in most cases even the investors. They all need to be kept informed of the progress on a regular basis. In general, a problem that surfaces early in the program gives stakeholders an opportunity to adjust their expectations and provide helpful feedback. As in life, nobody likes surprises especially the nasty ones. Therefore, any negative news about the project should be communicated sooner rather than later. If a project was deemed to be on time, then stakeholders may plan other activities based on the availability of this product. As an example, let us consider procurement of ambulances for a healthcare facility. With a promising level of progress, the stakeholders may prefer to advance the date of procuring more expensive and better equipped ambulances with ultrasound devices communicating via real-time networks with the hospital's Cloud support.

As seen previously in Fig. 7.2, it is important to maintain a low latency between the ultrasound machine's measurements reaching Cloud first and then the signals reaching users' mobile devices to make timely decisions. A Program Manager needs to outlay goals for an acceptable level of latency for the design team. It is not possible to have zero delays, so a reasonable goal would be less than 250 ms, which is the time between two sequential blinks of an eye. This response time is often targeted for real-time responses over the Internet.

As for the tools to measure round-trip network latency, one may consider using Ping, iperf, or netperf. However, these tools are not all implemented and configured in the same way. Different tools may return different set of results. So, the tools may be selected based on the observations that are to be monitored or measured. A highly recommended tool is Ping, which sends one ICMP (Internet Control Message Protocol) packet to the specified IP address and measures the Round-Trip Time (RTT) for transactions.

Performance testing invariably requires involvement of a beta customer site for realistic measurements. If such a Cloud-based application is running late in its product delivery, then the hospital may choose to delay the purchase of new ambulances or hiring additional personnel. To avoid disappointments later, it is important to have some built-in time buffer (referred sometimes as slack) for unforeseen problems. However, a large buffer is not good either. An unexpected early delivery is as much of a problem as a late delivery. This means that medical devices or applications will be idling or waiting for users to get ready. It also means that some development resources could have been saved or used elsewhere.

A project is deemed successful in the business world if it is delivered on time, perhaps within a margin of 10% of the original schedule, with high quality full features as planned with complete documentation. Since we mentioned SAMD (Software as a Medical Device) application, this may need a separate regulatory clearance, such as 510k from the FDA (Food and Drug Administration) in the USA, or similar regulations as applicable elsewhere. The time required to obtain the FDA approval should be accounted for in the project plans.

7.8 Across Time-Zone Management

If a project's execution team is split across multiple time zones, then there will be inefficiencies in communication as a daytime in the USA will be night in a location like India. Thus, additional time should be accounted for issues that require real-time debug meetings with participants from both sides of the ocean. A non-overlapping time zone shrinks the working hours to relatively fewer as compared to when both teams are available within a single time zone. Furthermore, one or the other side of the team is always making a sacrifice of their evening, nights, or early morning hours to deliver this project. This by itself takes a toll on morale and team spirits.

For remote team members to work and collaborate on a common code base, one option is to use Gihub [7]. It allows repositories to be setup with version controls, check in and check out facilities. A repository helps to organize a single project and keep track of its progress, as shown in Fig. 7.5. Repositories can contain folders and files, images, videos, spreadsheets, and datasets — anything the project needs. Often, repositories include a README file, a file with information about the project.

By default, a code repository has one branch named Main, which is the definitive branch for accessing project-related entities. One can create additional branches emanating from Main in the repository. These branches can maintain different versions of a project at any given time. This is helpful in adding new features to a project without changing the Main source of code. The work done on different branches will not show up on the Main branch until one merges it. Branches are used to experiment and make edits before committing them to Main. Figure 7.6 shows the Main branch, a new branch called feature and the journey that feature takes before it is merged into Main.

For tracking issues, such as bugs, enhancement requests, etc., Jira [8] can be used. It enables us to plan, track, and manage Agile software development projects. Stakeholders can create tickets and assign to the development or validation team members as required.

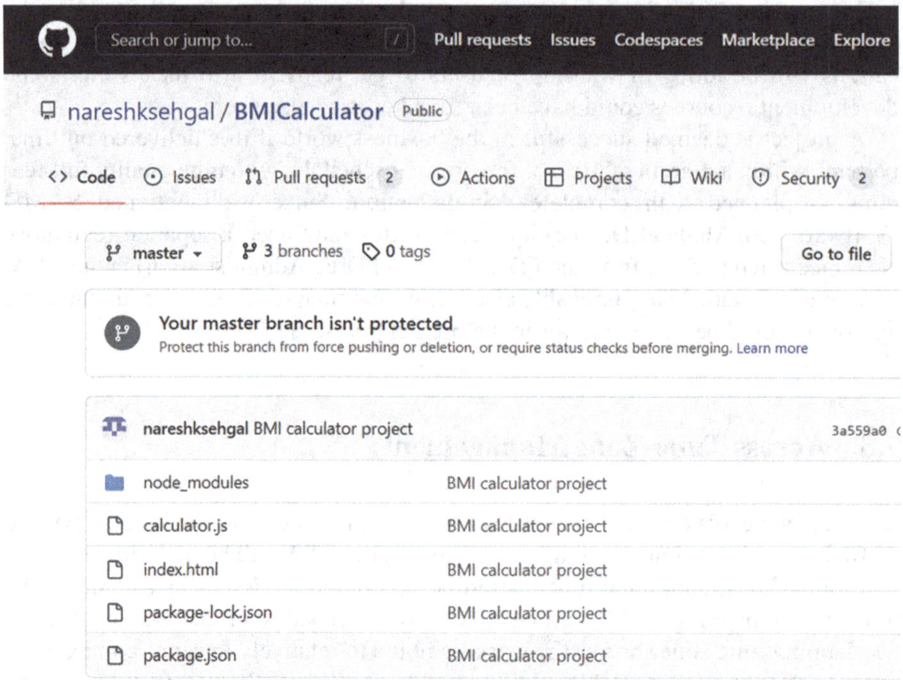

Fig. 7.5 A screenshot of code repository in Github [7]

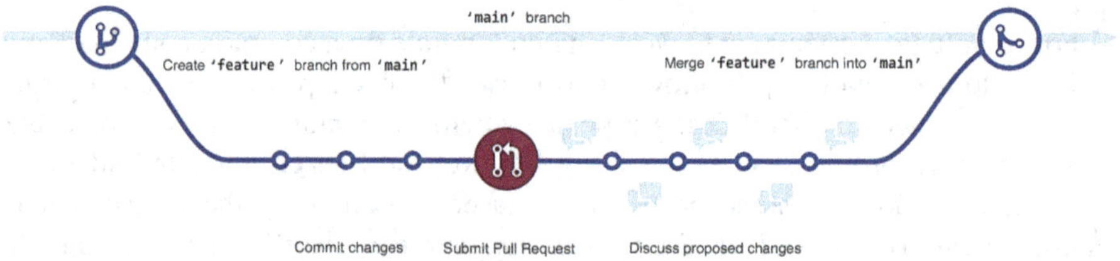

Fig. 7.6 Splitting and merging a branch to a codebase in Gihub [7]

A dashboard in Jira software is a part of a project that displays issues giving teams a flexible way to view, manage, and report on work in progress. Figure 7.7 is a visual representation of a team's dashboard for a project.

Thus, the obstacle of time zone difference can in fact be turned into an advantage. For example, if a bug is found by US team members during daytime and entered as a Jira ticket, then before the next morning time in the USA developers in India can review and fix the bug. This enables round-the-clock development and testing cycle across the globe.

During the deployment phase of the application, Cloud portal needs to be carefully monitored for the latency of data traversal from devices in a hospital to the datacenter, and then access times from mobile devices to the Cloud portal. As the

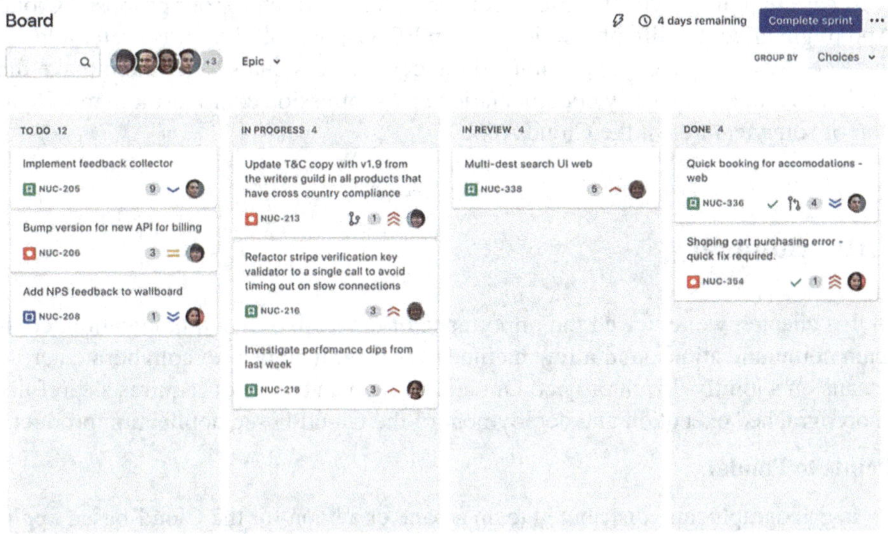

Fig. 7.7 An issue tracking board in Jira [8]

number of users and patient examinations grow over time, some extent of advance planning is needed to ensure that the SLA can be met. This needs Cloud-based scalability. A detailed discussion for which can be found in [9].

7.9 Regulatory Documentation

Since the Cloud App described here was intended to be used for a medical purpose, it needed approvals by a regulatory body such as the FDA. In the USA, FDA has recently ruled [10] that any software which extends beyond medical device hardware needs an explicit approval before deployment as SaMD (Software as a Medical Device).

Approval by FDA requires extensive documentation starting from the user needs, product requirements, architecture, implementation details, list of tests and their outcome etc. It is better to start and complete these documents as the project progresses through different phases of requirements gathering, design and testing, etc. This enables the management to make a pre-submission of SaMD application to FDA, which is highly recommended. This provides an opportunity to start a dialog with the regulators before Cloud App is ready for the market. The benefits of an early submission can be enormous. It is a chance to receive formal feedback from the FDA and get their stance on various issues in writing. This also eliminates a lot of guesswork moving forward, as product managers can refer to this feedback at the time of making decisions about the product.

A case in point is AI/ML-based decision-making, which for our proposed Cloud App might affect the life of a patient. From FDA's point of view, it is preferable for the App to give hints or suggestions to medical professionals to let them make the final decision. Hence, early documentation and submission could result in modification of some features of the Cloud App.

7.10 Summary

In this chapter, we reviewed the importance of device-based communication, cross team communications, and a few methods for team members to collaborate across oceans on a jointly driven project. Obviously, such an endeavor requires a carefully choreographed execution and deployment of the Cloud-based application product.

Points to Ponder

1. Is a geographically distributed team a bane or a boon for the Cloud-based application development project?
2. What are some of the methods to keep project stakeholders motivated?
3. How can latency be minimized for the Cloud-based application's users?
4. What benefits can data-driven project management provide?

Answers

1. Is a geographically distributed team a bane or a boon for the Cloud-based application development project?
   (a) Generally, for hardware projects, it is desirable to co-locate the implementation team to enable sharing of ideas and components in a lab.
   (b) However, for software projects especially the ones using Internet and Web-based components, it is possible to leverage the round-the-clock work hours available across time zones so that the developers at one location can work on fixing the bugs found by validators at the other locations. This requires use of common tools such as Github [7] and Jira [8].
2. What are some of the methods to keep project stakeholders motivated?
   (a) A Cloud-based application project may span many years, so features should be categorized into cogently clustered in the releases. Each release can be made periodically—say every 6–9 months.
   (b) For each product release, the project manager can plan a launch event where the work of the team members can be recognized. This gives the team a brief break to celebrate and avoid fatigue from working on long duration projects.
   (c) It is important for the project manager to invite higher management and executives, including from the user community to the project launch celebrations.

3. How can latency be minimized for the Cloud-based application's users?
   (a) Since data to and from a Public Cloud travels across several intermediate Internet nodes, some latency is inevitable.
   (b) Latency can be minimized by choosing a Cloud portal closer to the data [9].
   (c) It is important to consider a hot swap datacenter location, to ensure continued availability of mission critical databases and application services.
4. What benefits can data-driven project management provide?
   (a) Applying data driven project management can improve the chances for success of the ongoing projects and increase predictability in the future projects.
   (b) Data-driven project management allows for accurate tracking of important metrics in projects, including time tracking, resource availability and sharing, scope creep, etc.
   (c) IoT devices can be used in data-driven project management to provide real-time data on workers' progress and safety.

References

1. https://www.novasignal.com/support/security
2. https://www.rapidai.com/
3. Bhatt, P. C. P. (2021). *Software design, architecture and engineering: Concepts and practice*. Prentice Hall India, Easter Economy Edition. ISBN-978-81-951611-4-0.
4. Mounir, E. K., Ahmed, A., Abdulla, A., Hamad, A., Meera, M., & Ahmed, A. (2023). Implementing IOT in effective project management. *International Journal of Computers and Their Applications, 30*, 192–200.
5. https://lightbug.io/
6. https://www.aatmunn.com/applications/safety-control-center
7. https://github.com/
8. https://www.atlassian.com/software/jira
9. Sehgal, N. K., Bhatt, P. C. P., & Acken, J. M. (2020). *Cloud computing with security and scalability*. Springer. https://link.springer.com/book/10.1007/978-3-031-07242-0
10. https://www.fda.gov/medical-devices/digital-health-center-excellence/software-medical-device-SaMD

Chapter 8
IoT Product Deployment, Maintenance, and Updates

8.1 Introduction

After an IoT product has been designed, manufactured, tested, and released, its journey into the real-world begins. While software products with online maintenance are often straightforward to deploy to a customer, IoT products include hardware elements. The lifetime of the product begins as soon as it leaves the manufacturer's warehouse and will include storage, shipping, initial setup, network connection, standard operation, and maintenance. Creating products that are easy to use and support over a long lifetime increases the value proposition for all parties involved. The aim of this chapter is to provide insight into all stages of the serviceable lifetime of an IoT product for development and management teams.

8.2 Hardware Quality and Levels of Concern

Developers of IoT hardware should choose an appropriate level of quality for their products. Some factors included in this choice are as follows:

- Desired usable product life for the customer.
- Environmental conditions for operation of the product.
- Ramifications of product failure.
- Time to market.
- Cost to manufacture, maintain, and update.
- Customer satisfaction.

In some cases, such as safety critical IoT products like smoke sensors or carbon monoxide sensors, the manufacturer is required to meet the highest quality standards. In other cases, the manufacturer has discretion to perform a systems analysis and design in the appropriate quality to meet business and customer needs.

P. C. P. Bhatt et al., *Project Management in Cloud and IoT Applications*, https://doi.org/10.1007/978-3-031-99427-2_8

Table 8.1 IEC62304 software classifications

Software category	Impact
Class A	No injury or damage to health is possible
Class B	No serious injury is possible
Class C	Death or serious injury is possible

Prior to 2023 for medical device software, the FDA defined the concept of a Level of Concern in the Guidance for the Content of Premarket Submissions for Software Contained in Medical Devices. The Level of Concern is an estimate of the severity of injury that a device could inflict, either directly or indirectly, on a patient or operator as a result of device failures, design flaws, or simply by employing the device for its intended use [1]. The three Levels of Concern were Major, Moderate, and Minor and relate to the quality process and amount of documentation needed for approval. As of 2023, the FDA has simplified this to a two-category system [2]; however, IEC 62304—Medical device software—Software life cycle processes [3] still maintains a three-category software classification (Table 8.1):

The above guidance and safety classifications are for medical device software, but it is useful to take a similar approach towards levels of concern for IoT. Levels of concern can be considered for all business-critical aspects of the IoT product (safety, device longevity, customer satisfaction, etc.). Creating a three-classification system can simplify design decisions when it comes to hardware quality. Below are two examples for illustration. If an organization or IoT product team chooses an approach like this, the choice of categories and impact should be appropriately tailored.

8.2.1 Example: Ramifications of Product Failure

See Table 8.2.

Minor Impact When a product's failure is not associated with safety, the design team need not heavily weigh electrical safety as a constraint. It is important to remember that IoT hardware contains electronics, and there is always some associated level of electrical energy in the device. If that level is low, which may be the case with a small, battery-operated device, then the concern remains minor. It may be possible to save design time and cost by not integrating multiple redundant, fault-tolerant circuits in the product. If the product design meets the applicable safety standards for its application, then the design and management teams are free to optimize other aspects of the IoT product.

Moderate Impact The middle class of a three-class system is easiest to describe as somewhere in-between the first and third class. For an IoT product where some non-serious harm can occur, the product team should design in the appropriate hardware safeguards to minimize the chance of hazards. An example is placing a physical hardware barrier around the drill or blade of an Internet connected milling machine.

Table 8.2 Hardware categories when considering product failure

Hardware category	Impact
Minor impact	No injury or physical damage to customers is possible. Inconvenience to customer only
Moderate impact	No serious injury is possible. Failure may significantly disrupt a customer's operations
Major impact	Death or serious injury is possible. Customer relies on the product and a failure causes a major disruption to life and operations

Table 8.3 IEC62304 software classifications

Hardware category	Impact
Short life	The customer considers the device replaceable and/or is satisfied if the device works for a reasonable period of time
Moderate life	The customer expects a long lifetime but understands that the unit or accessories such as batteries, gaskets, or filters may need to be replaced from time to time
Long life	The customer only expects to purchase the device a few times in their life and that little or no supervision is required

Major Impact When a person's life is in danger due to failure of an IoT product's hardware, the bar for quality and safety should be the highest. The product team should consider "intrinsic safety" in their electronics design. Intrinsic safety is a protection method for electronics which operate in hazardous areas [4]. Consider a harmful gas detector in a coal mine. If the electronics create a spark or fire, there could be a major disaster. Intrinsic safety places limits on the available electrical energy in a product and ensures safeguards are in place (such as current limiting resistors or voltage limiting Zener diodes) to minimize the effect of any device failure.

8.2.2 Example: Desired Usable Product Life for the Customer

See Table 8.3.

Short Life Examples of electronics in this category may include personal gadgets like Bluetooth headphones, thumb drives, or smart charging devices. For products in this category, it is reasonable to assume that the customer expects changes in technology over time to change the form factor or capabilities of the device class. Power and data communication ports may change (e.g., USB-A to USB-C) or electronics innovations may evolve quickly (the increasing resolution of digital cameras over time). For products with a short lifetime, the most advanced, highest quality, and robust sensors need not be used. The device and components used need only keep up with the customer's expected pace of change.

Moderate Life Moderate life products could be smart-home appliances, cell phones, or entertainment products like smart TVs. These products are expected to have a long life but will reasonably be bought many times in a customer's life. This category gives design teams the flexibility to balance initial design time and cost of components against the upkeep cost of maintaining the product once deployed.

Long Life Some products may have very long-life expectations. An example is an Internet-connected automobile. For these products, incorporating high-quality, cutting-edge technology and robust components should be considered. If lower quality parts are used, the cost of maintaining the product throughout its lifetime may marginalize the profitability of the project.

It can often be unwise to underestimate customers' expectation of hardware product quality. Whether intentional as a profit-making strategy or unintentional due to a lack of consideration for hardware quality, the shorter-than-expected life of a poor-quality product may be categorized as "planned obsolescence." Planned obsolescence is achieved with techniques like designing in components known to wear down overtime or software bloat to slow down a product [5]. The goal of planned obsolescence is to make customers buy products again and again. It is often perceived as a scheme to wring more money out of customers and is viewed negatively. By designing IoT hardware with quality and levels of concern in mind, companies can provide great products and avoid the harmful image of planned obsolescence.

8.3 Packaging and Shipping

With real-world devices, a design area that is often overlooked or understaffed is packaging. Designing appropriate packaging or choosing the best off-the-shelf packaging for a product mitigates the risk of damage during shipping and provides for the best "unboxing" experience. As the first experience, a customer has with an IoT product is opening the box and initial setup, it may have an oversized impact of perception of the entire product. The systems and packaging team should consider the robustness of the packaging, the overall packaging waste, and the unboxing experience.

The International Safe Transit Association (ISTA) is a global alliance of shippers, carriers, testing laboratories, and more in the field of transport packaging [6]. Their mission is to empower organizations and their people to minimize product damage during the distribution and optimize resource use through effective package design. ISTA provides test procedures and test laboratories that can be used to gain confidence in a product's packaging. ISTA's test protocols are organized into series and are described in Fig. 8.1.

Figure 8.2 shows the test protocols recommended based on distribution and package type.

An IoT product team may use the information in Fig. 8.2 to plan their package design and test. For example, in the design phase of packaging for a typical IoT

1 Series: Non-Simulation Integrity Performance Tests.
Challenge the strength and robustness of the product and package combination. Not designed to simulate environmental occurrences. Useful as screening tests, particularly when used as a consistent benchmark over time.

2 Series: Partial Simulation Performance Tests.
Tests with at least one element of 3 Series type General Simulation performance tests, such as atmospheric conditioning or mode-shaped random vibration, in addition to basic elements of a 1 Series type Non-Simulation Integrity test.

3 Series: General Simulation Performance Tests.
Designed to provide a laboratory simulation of the general damage-producing motions, forces, conditions, and sequences of transport environments. Applicable across broad sets of circumstances, such as a variety of vehicle types and routes, or a varying number of handling exposures. Characteristics will include simple shaped random vibration, different drop heights applied to the sample package, and/or atmospheric conditioning.

4 Series: Enhanced Simulation Performance Tests.
General Simulation tests with at least one element of Focused Simulation, such as test sequence or test conditions linked to actual known distribution. Project 4AB is currently the only protocol in this Series. 4AB is a web-based Enhanced Simulation Test Plan generator, with on-line access available at no charge to all ISTA members. Project 4AB closely ties the tests and sequence to a user-defined pattern of distribution, and includes a broad range of current and quantitative information on distribution environment hazards. See the Project 4AB section of the ISTA® website (www.ista.org) for more details.

5 Series: Focused Simulation Guides **NOT AN ACTIVE SERIES**
Guides for the creation of laboratory simulations based on actual field-measured and observed hazards and levels. The 5-Series are not performance tests per se, but information and instructions related to establishment of user-defined Focused Simulation tests.

6 Series: Member Performance Tests.
Test protocols created by ISTA members or by ISTA in cooperation with a member company, to meet their particular purposes and applications. The tests may be completely original, or may be modifications or variations or ISTA Procedures or Projects or other published and accepted tests. The tests are reviewed, approved, and formally accepted by successfully completing ISTA's procedure for developing Test Protocols, per ISTA's Technical Division Operating Guide.

7 Series: Development Tests.
These tests are used in the development of transport packages. They can be used to compare relative performance of two or more container designs, but are not intended to evaluate the protection afforded packaged-products.

Fig. 8.1 ISTA Test Series [7]

consumer product weighing less than 68 kg, the 1A test protocol may be used. These tests do not simulate environmental conditions of shipping but do test the strength of packaging. Test protocols 2A and 3A increasingly simulate damage inducing stresses such as vibration, shock, and atmospheric conditions present during transport. These tests can be performed on the completed package and product design to provide the highest confidence that it will arrive at the customer's doorstep undamaged.

Reducing packaging waste has many advantages. First, eliminating extra packaging materials through efficient design reduces the overall BOM cost of the packaging and product. Second, less packaging can mean less material in landfills to the benefit of all. Third, customers may positively associate a company's appreciation of the environment with that company's brand. However, it is important to balance the positive aspects of sustainable design with the occasionally negative perception of quality with eco-friendly products [8]. Moulded plastic packaging inserts can be seen as higher quality than corrugated cardboard inserts, for example.

Developing a pleasant unboxing experience is a best practice that applies to all types of products, from clothing to IoT electronics. Many design firms offer advice

Distribution Type	Package Type		
	Individual Packages		Unitized / Pallet Load
	up to 150 lbs. (68 kg)	over 150 lbs. (68 kg)	
Parcel Delivery	3A, 6-FedEx-A	Not Applicable	Not Applicable
LTL (Less-Than-Truckload) Delivery	3B	3B, 6-FedEx-B	3B, 6-FedEx-B
Manufacturing to Distribution Center	Not Applicable	Not Applicable	3E
Distribution Center to Retail	3F	Not Applicable	Not Applicable
eCommerce: e-retailer Fulfillment	6-Amazon.com-Over Boxing, 6-Amazon.com-SIOC	6-Amazon.com-SIOC	Not Applicable
Club Store	6-Sams Club	6-Sams Club	6-Sams Club
European Consumer Goods	3K	Not Applicable	Not Applicable
Specialized Furniture	2C	2C	2C
Thermal Testing/ Testing of Insulated Shipping Containers	7D, 7E	7D, 7E	Situational
Bulk/ Automotive Resuable	Not Applicable	Not Applicable	3H
User Defined/ Custom Test Planning	4AB	4AB	4AB
Preliminary Design Screening/ Integrity Test	1A, 1C, 1G	1B, 1D, 1H	1E
Partial Simulation	2A	2B	Not Applicable

Fig. 8.2 Applicable ISTA test protocols [7]

on the unboxing experience, but global packaging consultant Innorhino offers common steps to include [9]:

- Know your audience—the unboxing experience should be tailored to the typical user of the product.
- Invest in Quality Packaging Materials—high-quality packaging materials reflect the value a design team places on its product.

- Incorporate Sustainable Packaging Solutions—as mentioned above, customers value eco-friendly packaging.
- Design for Shareability—photo and video sharing on social media can lead to viral, crowd-sourced marketing.

Product packaging is linked to the brand of an organization. For example, Apple's packaging represents luxury and minimalism [9]. When packaging and unboxing is given attention at the design phase, the beneficial payoff applies to both the product and company and can have a compound effect on profitability.

8.4 Initial Setup and Connecting IOT Products

When an IoT product arrives at its customer's premises receiving dock, it is ready for installation and immediate use. Ideally, there would be a simple sequence of steps leading to quick results for the user or organization employing the product. In a few cases, a more detailed initial setup is required even for commercial products. For industrial or infrastructure IoT networks, the setup may be complicated and require service and application engineer support from the manufacturer for success. The IoT project team should ensure their organization is ready to seamlessly support initial setup of their IoT products.

Figure 8.3 is a flow diagram for the typical setup of a commercial IoT product. Not all steps are necessary for all devices and industrial or infrastructure products may require a few more steps. However, this flow highlights some key points. For each step in the flow, the expected support from an IoT project team and manufacturer is listed.

8.4.1 Unbox

As mentioned above, unboxing is the first interaction a customer will have with a product.

Expectations from the IoT Project Team

- A pleasant unboxing experience (note it was considered in the design phase).
- An easy-to-use quick start guide or link to one is included in the package.
- Ideally, all accessories needed, such as a USB cable for charging, are included.
- All online instructional videos or guides should be deployed on the Internet, active and up to date.

Fig. 8.3 Typical setup flow for a commercial IoT product

8.4.2 Charge Device

Some IoT products will need to be charged prior or during their first use. This step is optional based on the design. Products that use primary (non-rechargeable) batteries will need the batteries installed after unboxing. Products that are not battery operated will need to be plugged into a power source. For products that use rechargeable batteries, initial power charging is necessary due to shipping regulations. For example, the International Air Transport Association recommends the state of charge for Lithium-Ion batteries not exceed 30% [10].

Expectations from the IoT Project Team

- The batteries included contain a required minimum level of charge.
- The product charges relatively quickly.
- The product can be turned on and is configurable while charging.
- The product provides a visual indication when charging.

8.4.3 Download App

In this context, the term "App" refers to that software which is not embedded in the IoT product but is used for interface or control. It could be the software that runs on a phone, tablet, or a progressive Web application that runs on any Internet browser. While not always required, an app is a common method to facilitate connecting the IoT product to the Internet. Instead of requiring a download, the app could be included inside the packaging on a portable media device (e.g., USB thumb stick). The app may require the creation of a personal account to be used, but oftentimes a user may prefer not to have to create an account to use a product.

Expectations from the IoT Project Team

- The app is available for download or included in the product packaging.
- If using a phone/tablet app, it is supported in both the Android and iPhone ecosystems.
- The app is up to date with maintenance releases.
- If support for the app is coming to an end, then pro-actively communicate this to the customer.
- The app is easy and quick to use.

8.4.4 Connect Device

Connecting to the Internet is typically required for fully initializing an IoT product. Some devices connect directly to the Internet, while others require a "gateway device" to connect to the internet. A gateway device could be a cell phone. The IoT product might pair to the cell phone via Bluetooth, at which point the cell phone acts as a gateway to the Internet for the IoT product. Connecting to the Internet may not always be required for initial setup but would provide access to most of the functions.

Expectations from the IoT Project Team

- Common methods for connecting to the Internet are supported (Wi-Fi, Cellular, etc.)
- Common methods for pairing to a gateway device are supported (Bluetooth).
- The product supports use on encrypted networks.
- The product indicates a successful connection to the internet.

8.4.5 *Personalize Device*

Device customization focuses on features that are meant to simplify or personalize the IoT product to the user. Examples are user interface settings like preferred graphics for a personal fitness tracker display or functional settings like the desired room temperature set point for a smart home thermostat. If the IoT product collects and transmits data to the Internet, the frequency of syncing that data is another common customization.

Expectations from the IoT Project Team

- Provide easy access to product settings, including privacy settings.
- Ensure all appropriate and meaningful parameters are configurable in the device settings.
- Maintain customization based on user profiles.
- Provide a method to backup customization settings.

Throughout all steps of the IoT product setup, there is an additional expectation that a customer support representative is available if needed. The method of customer support will depend on the business goals, organization, and complexity of the product. The representative may be virtual, as with an AI chat bot, or a traditional support technician available by phone.

8.5 Offline Use of IOT Products

Although commonplace, Internet connectivity is not always guaranteed. Portable devices may enter Internet dead-zones. Fixed location devices may experience Internet outages. In all situations, customers often expect their IoT products to maintain some basic level of functionality. In some cases, such as in safety-critical healthcare, functionality may be required regardless of access to the Internet. IoT product design teams should keep the following design principals in mind to minimize the impact due to the loss of Internet connectivity. Each of these will be illustrated with one good design example and one poor design example.

- Critical functionality should be embedded and always operational.
- Products should provide some valuable amount of utility even when not connected to the Internet.
- Loss and restoration of Internet service should not be overly disruptive to the user experience.

A key capability to IoT products is their inherent embedded processing. When working in tandem with the Cloud, the overall processing power available can be astounding. However, it is important for an IoT product not to be overwhelmingly dependent on an Internet connection to perform its most critical tasks. This is especially true for safety-critical applications like healthcare or transport. Consider the design of a collision avoidance IoT subsystem of a self-driving car. The subsystem processes camera images and relies on image processing to uncover potential collision hazards and communicate those hazards to the main processor of the car. The collision avoidance subsystem continually improves its object detection through Cloud-based machine learning. To do this, the subsystem needs Internet connectivity and bandwidth to transmit data. A poor design example would be to require the subsystem to transmit data to the Cloud and rely on results from the Cloud in real-time to detect objects. Doing so would impair the detection ability when no Internet connection is present and may be too slow even with Internet if the connection has low bandwidth or is intermittent. A good design example would be to ensure all object detection processing occurs in real-time on processors embedded in the local hardware. Once real-time detection has been completed locally, the data and results can be cached and uploaded to the Cloud at a convenient time to improve future object detection.

The above principal applies to non-critical functionality as well. IoT products should always maintain some level of utility for their users. Doing so increases the value proposition for the product which may result in increased market share. The consideration when the functionality is non-critical, however, becomes a subjective value-based trade-off for the product team and the entire feature set of the IoT product. Imagine a smart energy meter for home use. The smart meter monitors total power consumption in a house throughout the day and can provide notifications for energy savings based on typical usage and grid costs. Data is also displayed to the user with a dashboard on the Internet. While all the features (power monitoring, predictive analytics, notifications, dashboard, etc.) could be implemented in a way that requires the Internet, this may be a poor choice. A better choice may be to choose some threshold for utility, such as basic power monitoring with an embedded display for offline viewing, which remains operational during an Internet outage.

Whatever breakdown between online-only and offline functionality is chosen, the user experience interacting with the product should always be considered. Users should not be forced to significantly modify their workflow when the

Internet is unavailable or intermittent. For example, a smart TV may attempt to connect and log into various TV content providers upon power-on. This allows the viewer to quickly start browsing TV. When the Internet is available, this process is seamless. However, a poor design choice would be to implement a lengthy and blocking connection retry process when the login process fails. In this scenario, a user may have no other option than to stare at a spinning wheel on their TV for minutes while it attempts to login to online services with no Internet available. A wise choice would be to quickly abandon the login effort if the Internet is unavailable and allow the user to quickly browse content from other input sources such as an antenna.

8.6 Customer Support

Customer support spans a wide range of commitments including training, troubleshooting, maintenance, and product end-of-life issues. An organization and project team should decide on the scope of customer support to be provided for a product. When offering customer support, the team should also consider providing a service-level agreement (SLA). An SLA is a contract between a service provider and a customer that lays out the expected level of service and resolution times. SLAs need not always be formal, but by providing customers with expectations on support, a baseline level of satisfaction can be established and promoted.

For IoT products, customer support will often include services related to hardware. As such, a manufacturer needs to have appropriate resources available to support and debug hardware issues. When an issue arises, it is possible and sometimes desirable for a customer to address the issue by themselves. For this, the project team should develop a simple-to-use troubleshooting manual or online portal. Common content for the manual includes a scheduled maintenance section and a simple "symptom" and "solution" table. Figures 8.4 and 8.5 are examples taken from the manual for a Smart Wi-Fi enabled LG Washer/Dryer Combo [11].

In other cases, field service technicians may need to be employed or otherwise retained. Technicians should have internal documents describing the design and failure modes in detail along with detailed troubleshooting steps which may require skilled expertise and tools. The manufacturer should also provide support for the support team. In other words, the project team engineers may need to provide help debugging hardware issues. A project manager should understand that some amount of an engineer's time may be needed for product support.

Another aspect of IoT product customer support relates to mobile apps. App support directly impacts the size of the customer base. When designing a product with a mobile app, the project team must decide if they are supporting the iPhone ecosystem, the Android ecosystem, or other mobile application markets. Customer's will always hope and assume that a product will work with their mobile phone, but the

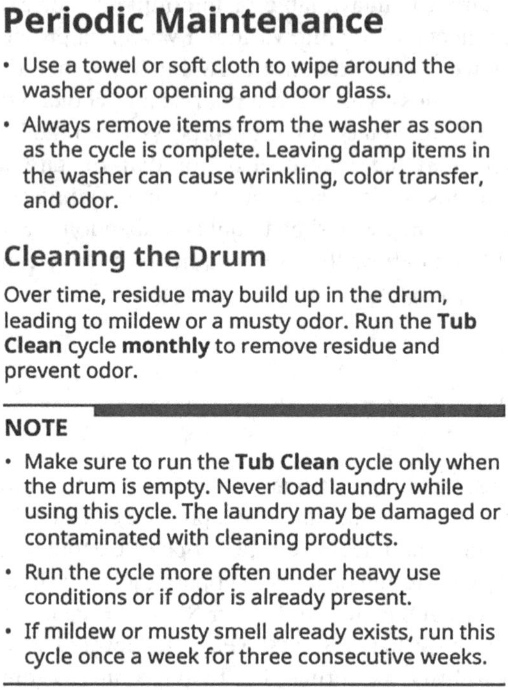

Periodic Maintenance

- Use a towel or soft cloth to wipe around the washer door opening and door glass.
- Always remove items from the washer as soon as the cycle is complete. Leaving damp items in the washer can cause wrinkling, color transfer, and odor.

Cleaning the Drum

Over time, residue may build up in the drum, leading to mildew or a musty odor. Run the **Tub Clean** cycle **monthly** to remove residue and prevent odor.

NOTE

- Make sure to run the **Tub Clean** cycle only when the drum is empty. Never load laundry while using this cycle. The laundry may be damaged or contaminated with cleaning products.
- Run the cycle more often under heavy use conditions or if odor is already present.
- If mildew or musty smell already exists, run this cycle once a week for three consecutive weeks.

Fig. 8.4 Example of periodic maintenance section in product manual [11]

design teams should plan and only commit to mobile apps they can support. If a product is introduced with support for both Android and the iPhone, but later in the product's lifetime support for Android is removed, customers will have a right to be upset.

With better customer support, higher customer satisfaction can be expected. Tips and tools for Design for Service are provided in Chap. 11. As many IoT products incorporate sensors, they are in a good position to collect data their own wellbeing in the background. Proactive data mining in the Cloud can help predict services issues with IoT products before they start.

44 TROUBLESHOOTING

Problem	Possible Cause & Solution
UE **UNBALANCE ERROR**	**The load is too small.** • Add additional items to allow the washer to balance the load.
	Heavy articles are mixed with lighter items. • Always try to wash articles of somewhat similar weight to allow the washer to evenly distribute the weight of the load for spinning.
	The load is out of balance. • Manually redistribute the load if articles have become tangled, preventing the automatic rebalancing from working properly.
dE, dE I, dE2, dE4 **DOOR OPEN ERROR**	**The door is not secured.** • Close and secure the door. If the message keeps appearing, unplug the power cord and call for service.
OE **WATER OUTLET ERROR**	**Drain hose is kinked, pinched, or clogged.** • Make sure that the drain hose is free of clogs or kinks and is not pinched behind or under the washer.
	The end of the drain hose is more than 96 inches (2.4 m) above the bottom of the washer. • Move the end of the drain hose so that it is lower than **96 inches (2.4 m).**
tE **TEMPERATURE SENSOR ERROR**	**Control error.** • Unplug the washer and wait 60 minutes. Reconnect power and try again. • If the error code is still displayed, call for service
FE **OVERFLOW ERROR**	**Water level is too high due to a faulty water valve.** • Close the water faucets, unplug the washer, and call for service.
PE **PRESSURE SENSOR ERROR**	**The water level sensor is not working correctly.** • Close the water faucets, unplug the washer, and call for service.
LE **MOTOR LOCKED ERROR**	**A motor error has occurred.** • Allow the washer to stand for 30 minutes and then restart the cycle. • If the error code is still displayed, call for service.
FF **FROZEN ERROR**	**Inlet hose, drain hose or drain pump is frozen.** • See **Care in Cold Climates.**

Fig. 8.5 Example of troubleshooting section in product manual [11]

8.7 Software Updates in the Field

Maintaining software and firmware embedded in IoT products is important to ensure a product's stability, security, and market competitiveness. Stability is addressed with bug fixes and platform enhancements. Security patches allow the product to remain at the leading edge of the battle between hackers and developers. Adding new features with software updates, whether user requested or driven from market research, can lead to a higher level of customer satisfaction.

The following provides a description of software update methods and a set of best practices. A more detailed description can be found in Stormotion's work entitled "How to Update IoT Devices: A Complete Guide to Secure Firmware Updates" [12].

8.7.1 Methods

- **Manual**—The manual software update method relies on somehow to physically connect to an IoT product with a storage media device such as a USB drive or SD card. Software updates are stored on the media, and a software update is initiated by the end user.
- **Over the Air (OTA)**—This method relies on wireless transmission of a software update package. The package could be transmitted via Bluetooth, Wi-Fi, or cellular. OTA updates can be initiated and performed remotely.
- **Hybrid**—A hybrid approach can combine manual methods with OTA methods. For example, a portable gateway device (such as a cell phone) to connect to the IoT product with Bluetooth. The gateway could then transfer software updates from the Cloud with a cellular connection.

Some pros and cons for the update methods are illustrated in Fig. 8.6 [12].

8.7.2 Business Best Practices [12]

- **Staged rollouts**—It is often beneficial to roll out new software to a small group of users first. This helps in identifying if there are any issues during or following the updates. It minimizes the impact of any update failures or newly introduced

Method	Pros	Cons
Manual Updates (USB, SD card)	Full control over the update process No reliance on the network connectivity	Not scalable — requires physical access to each device High downtime, labor-intensive, and costly at scale
OTA Updates via BLE/Wi-Fi	Highly scalable — enables mass updates remotely Minimal user friction, no physical intervention needed	Connectivity issues — BLE stability challenges and potential dropouts Rollback challenges — improper handling can brick devices
Hybrid Approach (BLE + Wi-Fi)	Combines local and remote updates for greater flexibility BLE ensures close-range updates, Wi-Fi handles larger firmware payloads	More complex implementation — requires dual communication protocols Higher power consumption, especially for battery-operated devices

Fig. 8.6 Pros and Cons for IoT software update methods [12]

issues. Careful monitoring of staged rollouts can pave the path for a smooth full-scale rollout.

- **Incremental updates**—Businesses can mitigate the risk of update failures by limiting and encapsulating the scope of changes. If each new software package only updates targeted modules, the chances of failure are reduced. This process can be done in conjunction with staged rollouts.
- **Fail-safe mechanisms**—In the event of an update failure, end users may desire to revert to a well-defined set point which brings the system to a known and functional state of the product. By enabling this ability, a manufacturer can reduce their customer support demands and avoid user frustration.
- **State of the art encryption**—Using strong encryption on software update packages reduces the risk of unwanted intellectual property exposure and unauthorized IoT product access or denial of service attacks.
- **Silent background updates**—Users may not prefer to be actively involved in software updates. Instead, they may prefer their device to have minimum downtime. Performing updates in the background at off-nominal usage times can enhance the user experience.

8.7.3 Technical Best Practices [12]

- **Adaptive Bluetooth Low Energy (BLE) transmission**—Given the chance for signal interference, BLE connections can have reduced fidelity. This may lead to incorrect or incomplete data transfer during updates. By adapting to environmental factors such as BLE signal strength and noise factor, a software update process can respond in a way which increases the likelihood of success.
- **Error recovery and retry logic**—Logic can be implemented to mitigate the risks involved with an interruption in the software update process. For example, updates can be resumed at the point when an interruption occurred. Alternatively, any interrupted updates can be restarted from a known state.
- **Multiprotocol support**—Allowing data transfer over multiple protocols can lead to a more robust update process. For example, if an IoT product has both Wi-Fi and cellular communications, it could support switching between those two protocols during update based on network availability.
- **Modular firmware architecture**—At the time of updating, a modular architecture that separates core from the non-critical functionality can be easier to manage. Also, by isolating the platform, security and connectivity modules in place, feature updates may present less risk overall.
- **Update scheduling**—Designs that consider the timing of software updates employ update scheduling. For example, a product may only allow updates to occur when a secondary power source is present. This would be the case for a battery-operated device that is also connected to the wall outlet. Another example is to schedule updates only when a product is in a low-usage period, minimizing the overall power consumption.

It is important for the IoT product teams to consider a thoughtful approach to software updates in the field. It is useful to understand the challenges involved in updating software on Internet connected devices. By applying some of the best practices described, teams can mitigate risk and improve the customer's experience.

8.8 Hardware Updates in the Field

Whereas software updates to IoT products are often expected and appreciated by customers, hardware updates to an already deployed device or product may be seen as a burden based on the nature of the application. This is because they inherently require disruption in the use of the product. With Commercial IoT, this can be frustrating. For example, the electronics on an Internet connected light switch may not be replaceable without removing the light switch from the wall or turning off a fuse. Circuit boards also generally require careful handling and can be damaged during replacement. Hardware updates might need to be done by specially trained service technicians with access to the product. The scheduling and in-person requirement could present a big inconvenience for both the customer and manufacturer.

With Industrial and Infrastructure IoT, however, hardware updates may be very welcome. These IoT components are expected to have more durability and longer life due to difficulty of upgrades. This is due to the particular use case and expectation of long-term functionality, initial installation, and product cost. Examples include smart parking meters or traffic cameras, which may benefit from hardware upgrades. A parking meter could be upgraded by adding solar cells to recharge internal batteries. As a city grows resulting in more cars on the roads, traffic cameras could be upgraded with higher bandwidth wireless hardware. With the above in mind, some methods and best practices are presented for consideration.

8.8.1 Methods

- **Adding hardware**—In this method, new hardware is provided that can be easily connected to the original hardware to expand functionality. An IO expander device that plugs into one USB port but provides many additional USB and multimedia card ports is an example.
- **Swapping parts**—Hardware can be replaced at a part or module level. A new enclosure can replace an out-dated enclosure. New circuit boards can be inserted in place of old circuitry.
- **Complete replacement**—Sometimes the only method available to update the hardware is to provide a complete product replacement. This can be the case when a product is particularly difficult to open or expand by design or when the cost for replacing the product is less than the cost to have it serviced.

8.8.2 Business Best Practices

- **Understand customer expectations**—The needs to support hardware updates are heavily influenced by the end user and application. Even in Commercial IoT, some customers enjoy tinkering with hardware and may not mind performing updates. Others may simply want a plug and play product that they never need to supervise. Understanding expectations upfront allows for efficient project management.
- **Factor in cost of upgrades**—If an IoT product is expected to support hardware upgrades in the field, it is beneficial to do a cost-benefit analysis on the various upgrade methods versus the hardware architecture. Investing in a design that easily supports upgrades could cost more upfront but pay off during the product's lifetime.
- **Provide customer service**—Appropriately staffing a customer service team is important for products that may be upgraded in the field. This team can be most cost-effective in the field and help to enhance customer satisfaction.

8.8.3 Technical Best Practices

- **Future-proof design**—Where possible, consider using a hardware platform that has resources and peripherals exceeding the initial design requirements. Early releases of a product may underutilize this platform, but the margins allow for seamless upgrades using the hardware already built into the system.
- **Modular design**—Modular designs have the benefit of supporting hardware upgrades in the field. For example, an Internet-controlled drone could be designed in such a way that its motors and blades can be replaced without affecting the main chassis. In this way, the drone can be re-configured to lift heavier loads or to operate in different atmospheric conditions. This contrasts with a unibody design where no easy modifications are possible.
- **Support for multiple hardware interfaces**—Hardware interface modules can be designed to support a variety of analog or digital protocols. For example, an end user may wish to control a television, air conditioning unit, and digital streaming device with a single universal remote controller. To do so, the controller's hardware may need to support infrared, radio frequency, and Wi-Fi transmission. The hardware need not be upgraded when interfacing with new devices. Ideally, a modular design and support for multiple hardware interfaces are incorporated.
- **Design for service**—Implementing design choices that simplify service can reduce the costs involved with customer support. By considering those who service and maintain a product as product users, scenarios are generated that inform the designers to satisfy their needs. The Cloud is particularly useful as a repository for service instructions, logs and device usage statistics. Regardless of the method used, hardware upgrades benefit from an architecture that considers Design for Service.

8.9 Summary

In this chapter, we cover both the design and operational management of IoT products. An emphasis is placed on hardware quality and other aspects such as packaging and shipping. A process of initial device setup, from unboxing to Internet connection and to affect personalization is described. Offline use of IoT products and customer support are considered as well. The chapter concludes with a set of best practices for software and hardware updates in the field.

Points to Ponder

1. What are some of the considerations when deciding the level of quality to design into the hardware for an IoT product?
2. Why is it important for development teams to consider the design of packaging materials used to ship the product?
3. What steps are necessary to operate an IoT product after it is delivered?
4. Why is offline use an important feature in IoT?
5. How is the customer service for IoT products different from software products?
6. What are the differences between updating software in the field versus hardware?

Answers

1. What are some of the considerations when deciding the level of quality to design into the hardware for an IoT product?

   (a) Product lifetime: The longer a product is expected to last, the higher the quality level is needed to support the usable lifetime.
   (b) Environmental conditions: If operating in an exposed or harsh environment, a higher level of quality can improve the chances of the device remaining operational when in use. For example, smart tractors operating in remote farm fields may experience extreme weather conditions during use and storage. This also applies to geo-tagged devices which are in remote locations.
   (c) Impact of failure: Safety and mission-critical products, such as telemetry sensors in satellites, can make the difference between a successful deployment and a failed operation. The impact of failure with these types of products typically drives a higher level of quality into the hardware design and validation.
   (d) Other factors: These include time to market, business costs, and customer satisfaction.

2. Why is it important for development teams to consider the design of packaging materials used to ship the product?

   (a) Functionally, the packaging protects a product against the damages which can occur due to shipping. Without proper packaging, newly manufactured devices can arrive in a non-functional state. This can lead to customer complaints and costly product returns.

   (b) Aesthetically, the packaging used reflects the quality of a product and the values of a company. Packaging design thus presents an opportunity to engage end users and support perceptions of company image.

3. What steps are necessary to operate an IoT product after it is delivered?

   (a) Unbox—Open the product and all associated accessories.
   (b) Charge (optional)—Products shipped with rechargeable batteries often need to be plugged in and charged before or during their first use. This ensures that the device will remain powered when proceeding through initial setup.
   (c) Download App (optional)—IoT products often use mobile apps or other software tools as a user interface. Some products may have embedded displays and controls which remove the need for an app.
   (d) Connect—An Internet connection is necessary to take full advantage of any IoT product. The product should support common methods to connect to the Internet like Wi-Fi and should be secure.
   (e) Personalize (optional)—End users may expect a certain level of customization to their needs. Adjusting device settings is often the last step of setting up a new product and these adjustments may change over time as a user's preferences change.

4. Why is offline use an important feature in IoT?

   (a) Customers are likely to expect their products to provide some basic level of functionality even if they are not online. In some IoT product environments, it may be well understood that Internet connectivity can be intermittent or weak. Products should always provide a certain minimum level of critical functionality in these cases.

5. How is the customer support for IoT products different from software products?

   (a) IoT products include hardware and that introduces unique customer service issues. Some hardware may require periodic maintenance by the user while other hardware may need the skills of professional service technicians.
   (b) Additionally, IoT products are often coupled with mobile apps or other software products. When the associated software is updated or new versions are released, care should be taken to ensure compatibility with IoT hardware. Removing support for associated software can result in limited functionality of the product.

6. What are the differences between updating software in the field versus hardware?

   (a) Expectations—Customers may expect new software features over time with products. They may also be more understanding if software updates are necessary to keep a device functional or even enhance and secure it better. For Commercial IoT, on the other hand, customers may prefer that the hardware they purchased does not require major updates. Industrial and Infrastructure IoT customers are more likely to tolerate or desire hardware updates.

(b) Difficulty—Generally speaking, it is more difficult to update hardware in the field than software. Hardware updates require physical access to a device. They also may require specialized training to ensure success.

References

1. https://www.fda.gov/files/medical%20devices/published/Guidance-for-the-Content-of-Premarket-Submissions-for-Software-Contained-in-Medical-Devices%2D%2D-Guidance-for-Industry-and-FDA-Staff.pdf
2. https://blog.johner-institute.com/iec-62304-medical-software/level-of-concern/
3. https://webstore.iec.ch/en/publication/22794
4. https://en.wikipedia.org/wiki/Intrinsic_safety
5. https://durabilitymatters.com/planned-obsolescence/
6. https://ista.org/
7. https://ista.org/docs/2020_ISTA_Guidelines.pdf
8. Ayebo, I. (2025). *Sustainable packaging and its effect on service quality.*
9. https://innorhino.com/blog/about-design/how-to-craft-an-unforgettable-unboxing-experience
10. https://www.lion.com/lion-news/october-2024/lithium-battery-30-state-of-charge-recommendations-for-2025
11. https://www.lg.com/us/support/product/lg-WM3555HWA.ABWEVUS?tab=1
12. https://stormotion.io/blog/updating-iot-devices/

Chapter 9
Regulatory Compliant Product Development

9.1 Introduction

In the USA, Food and Drug Administration (FDA) was established [1] to protect and address the public health issues. FDA ensures the safety, efficacy, and security of human and veterinary drugs, biological products and medical devices, etc. These regulations present a watermark that every medical company must satisfy before being allowed to ship its products. This requirement also covers software applications that extend the functionality and utilization of medical devices. Other countries follow similar regulations to protect their citizens. The medical industry is rife with regulations for good reasons. This must be seen and appreciated in the light of the fact that in the past some companies had developed products and brought them to market with insufficient testing causing undue harm to the patients [2, 3].

In addition to the above, additional rules and protocols exist about managing and protecting Personally Identifiable Information (PII) and Patient Health Information (PHI), such as CCPA [4], GDPR [5] and HIPPA [6]—just to name a few. These measures were well intended and put in place as a reaction to some adverse events in the past [7]. The medical regulations now create safety barriers to entry for the new healthcare products and services. This clearly increases the cost of product development and launches.

In this chapter, we shall also visit a different idea—in fact a new medical application, the opportunity for which emanates from the above regulations. Such an application will allow individual patients to collect and share their medical records across doctors, hospitals, and countries. Before discussing the App, we will examine additional challenges that applications dealing with PII and PHI data must overcome.

P. C. P. Bhatt et al., *Project Management in Cloud and IoT Applications*,
https://doi.org/10.1007/978-3-031-99427-2_9

9.2 IT Challenges

As elaborated in Chap. 5, information technology (IT) platforms provide the lifeblood of a company that curates and maintains medical data. In the past, companies stored their data on premises, but with the advent of Cloud Computing, it is often cheaper and more convenient to do all data processing in Cloud [8]. However, data needs to be moved from its generation point to a Cloud datacenter. If the data involves PII or PHI elements, then IT department of that organization would be expected to incorporate additional safeguards related to confidentiality, integrity, and availability of data in the Cloud.

For the Cloud application discussed in the previous chapter, a typical hospital will have hundreds of questions that need to be satisfactorily answered before a data pipeline to the Cloud datacenter can be set up. A sample scoring sheet is shown in Table 9.1, with over 150 questions that need to be individually answered. The scores in sections from A through K are checked to determine if the company should be permitted to move hospital data to the Cloud. Most of this data belongs to the patients. Patients may be permitted to access their own data directly. Imagine this scenario from a patient's point of view. She may have to visit multiple hospitals and doctors over her lifetime, sometimes for the progression of a single disease. In this case, doctors at each hospital may need to access this patient's records from the other hospitals but are unable to do so due to stringent data protection rules in place. Even though HIPPA [6] clearly states that patients have the right to their own data, the only way to access is by the old-fashioned paper-based request forms. Then several days later, data is copied to physically immutable portable storage media and handed over to the patient. Currently, there are no exceptions to this process. This is the case even for emergency cases, unless the hospitals are connected or affiliated with each other.

This situation clearly presents a new opportunity for patients to exercise their rights under the HIPPA [6] regime and move their own data to the Cloud without going through the previously described process, or completing the questionnaire shown in the Table 9.1. Then patients can continue to collect their medical data in a personalized and secure portal in the Cloud, which can be readily available whenever patient needs medical care elsewhere. We will describe the attributes and features of the proposed App in a later section in this chapter. First let us briefly review the nature of data management.

9.3 Data Management

Medical datasets have two aspects, one which can be used to identify personal information of the subjects from whom this data was collected, and the other which is related to the measured parameters. Researchers are often more interested in the latter to develop future drugs, medical treatments, and recommendations. They may wish to know the age and gender of the subjects, which are not considered PII or PHI.

Table 9.1 List of questions from a typical hospital before storing data in the Cloud

A. Company information. The vendor:
1. Will accommodate an onsite visit for a security audit within 24 h notice
2. Will store all confidential data within the USA—incl. backups
3. Maintains an audit log for the location of all confidential data and their backups, to identify where it is located at any point in time, in order to address privacy laws for storage within the USA
4. Will not access confidential data from outside of the USA
Total company controls
B. Policies, standards, and procedures. The vendor:
1. Has formal written information security policies
2. Will provide copies of the information security policies
3. Can provide results of a third-party external information security assessment conducted within the past 2 years (SOC Type 1 and 2, penetration test, vulnerability assessment, etc.)
4. Maintains incident response procedures
5. Has a policy to protect client information against unauthorized access; whether stored, printed, spoken, or transmitted
6. Has a policy that prohibits sharing of individual accounts and passwords
7. Has a policy that implements the following information security concepts: Need to know, least privilege and checks and balances
8. Requires system administrators to be educated and qualified
9. Implements AAA (Authentication, Authorization, Accounting) for all users
10. Performs background checks for individuals handling confidential information
11. Has termination or job transfer procedures that immediately protect unauthorized access to information
12. Provides customer support with escalation procedures
13. Has documented change control processes
14. Requires contractors, subcontractors, vendors, outsourcing ventures, or other external third-party contracts to comply with policies and customer agreements
15. Has a policy that implements federal and state regulatory requirements
16. Maintains a routine user information security awareness program
17. Has a formal routine information security risk management program for risk assessments and risk management
Total policy controls
C. Architecture. The vendor:
1. Will provide a network topology diagram/design of the technical solution
2. Implements network firewall protection
3. Implements Web application firewall protection
4. Implements host firewall protection
5. Maintains routers and ACLs
6. Provides network redundancy
7. Has IDS/IPS technology implemented
8. Uses DMZ architecture for Internet systems
9. Adheres to the practice that Web applications, which "face" the Internet, are on a server different from the one that contains the database
10. Uses enterprise virus protection on all systems

(continued)

Table 9.1 (continued)

11. Follows a program of enterprise patch management
12. Provides dedicated customer servers to segregate data from other customer data. If not, then how is this accomplished in a secure virtual or segmented configuration
13. Implements controls to restrict access to data from other customers
14. Ensures that remote access is only possible over secure connections
15. Uses separate physical and logical development, test and production environments and databases
16. Secures development and test environments using, at a minimum, equivalent security controls as the production environment
17. Will provide the architectural software solution design with security controls
18. Has managed, secure access points on its wireless network
Total architecture controls
D. Configurations. The vendor:
1. Implements encryption for confidential information being transmitted on external or Internet connections with a strength of at least AES 256 bit or uses TLS 1.1, preferably TLS 1.3
2. Implements encryption for confidential information at rest with a strength of at least AES 256 bit
3. Has password-protected screen savers that activate automatically to prevent unauthorized access when idle, for computers used by system's support users
4. Removes all unnecessary services from computers
5. Uses file integrity monitoring software on servers (such as tripwire, etc.)
6. Changes or disables all vendor-supplied default passwords or similar "published" access codes for all installed operating systems, database management systems, network devices, application packages, and any other commercially produced IT products
7. Uses passwords that are a min. of eight characters, expire at least annually, and have complexity requirements
8. Ensures that passwords are never stored in clear text or are easily decipherable
9. Checks all systems and software to determine whether appropriate security settings are enabled
10. Manages file and directory permissions following least privilege and need-to-know practices
11. Implements redundancy or high availability for critical functions
12. Authenticates all user access with either password, token or biometrics
13. Formally approves, tests and logs all system changes
14. Does not use production data for both development and testing, unless it has been declassified
15. Uses synthetic data in both development and test environments
16. Limits access to development and test environments to personnel with a need to know
17. Sets the account lockout feature for successive failed logon attempts on all system's support computers
18. Prohibits split tunneling when connecting to customer networks
Total configuration controls
E. Product design. The vendor:
1. Ensures that if the product integrates with portable devices, confidential information is encrypted when stored on these portable devices and requires password access

(continued)

Table 9.1 (continued)

2. Ensures that access to confidential information, across a public connection, is encrypted with a secured connection and requires user authentication
3. Implements protections for Common Vulnerabilities and Exposures (CVEs) in a timely manner to protect from exploits
4. Audits the application against the OWASP Top 10 Application Security Risks
5. Ensures that application server and database software technologies are kept up to date with the latest security patches
6. Uses threat modeling in their software development life cycle (SDLC)
7. Performs security code reviews as part of their SDLC
8. Conducts OWASP code reviews for the Top 9 source code flaw categories as part of their SDLC
Total product design controls
F. Compliance. The vendor:
1. Will provide relevant certificates of applicable ISO 27001 certification
2. Can provide documentation that its product is HIPAA compliant, if the vendor manages any PHI
3. Can provide documentation of its PCI-DSS compliance if the vendor manages any payment card information
4. Uses industry standard best practices for application security (e.g., OWASP)
Total product design controls
G. Access control. The vendor:
1. Immediately removes, or modifies access, when personnel terminate, transfer, or change job functions
2. Achieves individual accountability by assigning unique IDs and prohibiting password sharing
3. Ensures that critical data, or systems, are accessible by at least two trusted and authorized individuals, in order to limit having a single point of service failure
4. Ensures that users have the authority to only read or modify those programs, or data, which are needed to perform their duties
Total access controls
H. Monitoring. The vendor:
1. Reviews access permissions monthly for all server files, databases, applications, etc.
2. Implements system event logging on all servers and records at a minimum who, what, and when for all transactions
3. Reviews and analyzes after hours system accesses, at least monthly
4. Reviews system logs for failed logins, or failed access attempts monthly
5. Reviews and removes dormant accounts on systems at least monthly
6. Reviews Web server logs weekly for possible intrusion attempts and daily for significant changes in log file size as an indicator of compromise
7. Reviews network and firewall logs at least monthly
8. Reviews wireless access logs at least monthly
9. Performs scanning for rogue access points at least quarterly
10. Actively manages IDS/IPS systems and alert notifications have been implemented
11. Performs vulnerability scanning at least quarterly. This is a mandatory requirement for EMRs

(continued)

Table 9.1 (continued)

12. Performs penetration testing at least annually, if the vendor manages any PHI. This is a mandatory requirement for all EMRs
13. Checks routinely that password complexity is adhered to
Total monitoring controls
I. Physical security. The vendor:
1. Controls access to secure areas, e.g., key distribution management (both physical and electronic), paper/electronic logs, and monitoring of facility doors
2. Controls access to server rooms and follows least privilege and need-to-know practices for those facilities
3. Has special safeguards in place for computer rooms, e.g., cipher locks, restricted access, room access log, and card swipe access control
4. Shreds or incinerates printed confidential information
5. Prohibits or encrypts confidential information on laptops and mobile devices
6. Positions desktops, which display confidential information, in order to protect from unauthorized viewing
7. Escorts all visitors in computer rooms or server areas
8. Implements appropriate environmental controls, where possible, to manage equipment risks, e.g., fire safety, temperature, humidity, and battery backup
9. Has no external signage indicating the content or value of the server room or any room containing confidential customer information
10. Provides an export copy of all of the customer's data in a mutually agreed upon format at the end of the contract
11. Follows forensically secure data destruction processes for confidential data on hard drives, tapes, and removable media when it is no longer needed and at the end of the contract term
Total physical controls
J. Contingency. The vendor:
1. Has a written contingency plan for mission critical computing operations
2. Has emergency procedures and responsibilities documented and stored securely at multiple sites
3. Reviews and updates the contingency plan at least annually
4. Has identified computing services that must be provided within specified critical time frames, in case of a disaster
5. Has identified cross-functional dependencies, so as to determine how the failure in one system may negatively impact another one
9. Has written backup procedures and processes
7. Tests the integrity of backup media quarterly
8. Stores backup media in a secure manner and controls access
9. Maintains a documented and tested disaster recovery plan
10. Uses off-site storage and has documented retrieval procedures for backups
11. Password protects and encrypts all backups
12. Provides rapid access to backup data
13. Labels backup media appropriately to avoid errors or data exposure
Total contingency controls
K. Vendor's business associates
1. Confidentiality agreements have been signed before proprietary and/or confidential information is disclosed to the vendor's business associates

(continued)

Table 9.1 (continued)

2. Vendor's business associate contracts, or agreements, are in place and contain appropriate risk coverage for customer requirements
3. Vendor's business associates are aware of customer security policies and what is required of them
4. Vendor's business associate agreements document the agreed transfer of customer's data when the relationship terminates
Total business relationships controls
Total Third-Party Controls

The process to remove PHI from a dataset is called de-identification. A person's name, date of birth, and address information are removed or masked, or replaced with some generic data. This alone is not sufficient because if a person has repeated medical tests, their results need to be grouped together to assist in building medical history of that person. This can be easily generated securely by using a hash function, the inputs for which can be that person's PHI. Note that the output cannot be reversed to determine the inputs. This is because even a small change in input will result in a large change of the hashed token's value. Thus, all medical records with same hashed token can be unambiguously and securely grouped together in the Cloud ensuring that researchers can observe trends or changes in a patient's condition over time.

Such a de-identification process may be required even when a person is dealing with his own data, so that they can comfortably share it with other medical professionals to seek a second opinion. Once sufficient data from multiple patients is made available in the Cloud, it can be used for federated learning [8], and also be aggregated to perform data mining [9]. In general, data mining models have the following two kinds of tasks:

1. **Predictive:** The basic idea in this case is to make a prediction about values of data using known results found from different datasets. For instance, a sustained high blood pressure over time can portend a brain stroke or critical cardiac condition.
2. **Descriptive:** The basic idea here is to identify patterns or relationships in the data. Unlike the predictive model, a descriptive model serves to explore the properties of the data being examined. This is done not necessarily to predict any new properties but to find some interesting hidden data patterns. To unveil hidden patterns in a given dataset Clustering, Summarization, Association rules, and Sequence discoveries are often used. For instance, existing patients' data can be clustered and labeled as healthy, or not healthy in certain ways such as heart or brain disease patients. Then the parameters from a newly conducted examination can be compared with pre-existing clusters to see if there is a similarity or close enough match. In that case, the labels of matching clusters can be used by the doctors to quickly diagnose a new patient's condition.

For an individual patient, data mining offers a new avenue to have a good idea of one's own medical condition. This may come in handy if emergency medical treatment is required when one is traveling away from home. There are many cases of doctors administering a drug, with no or limited access to prior medical history, resulting in unintended side effects or reactions (e.g., to allergies). Furthermore, having access to one's own medical records enables a patient to get a second opinion before undergoing a major treatment regime.

Medical practitioners match a patient's blood test results with the established normal ranges to detect significant levels of deviations. Similar to blood test reports, a patient's overall attributes can be compared with other medical data in a de-identified dataset to uncover latent health characteristics. This can be done using clustering of patients by age, gender, and other non-PHI-based statistics in a large healthcare database. When a new result is uploaded, a normative range comparing this patient's results with other similar patients will be more meaningful. This requires Clustering, which is defined as an unsupervised learning to group data into similar groups or buckets. It may help with faster diagnosis of any potential issues. A medical problem caught early and diagnosed early paves the path for faster treatment.

However, clustering also presents some challenges, namely:

1. The right number of clusters is generally not known a priori. Also, handling the Outliers is difficult, as they may get placed in their own clusters.
2. There may not be any a priori knowledge concerning the clusters. Unlike learning during a classification process, where prior knowledge exists from the attributes of a dataset, in clustering there is no supervised learning to aid the process. This has caused regulatory agencies to not approve many AI/ML-based techniques for diagnostics purposes. At best, clustering can serve as an aid for a medical practitioner to arrive at an informed conclusion, but not for a patient to judge his own condition.
3. Cluster results are dynamic, as cluster membership may change over time.

Although a study of other available AI/ML techniques is beyond the scope of this book, more details can be found here [10] and are summarized in Fig. 9.1.

As shown in Fig. 9.2, each user can start sending private data for computations, after being authenticated into the system. Proxy server hides the traceability of messages sent from each user towards the Cloud server. This data is encrypted to provide protection and integrity against a man-in-the-middle attack. The analyzer in this model (as shown in Fig. 9.2) could be an external party, which extracts the statistical parameters of the user data by decrypting and performing analytics on it. There are obvious questions on the performance and efficiency of Cloud environments to deploy such a model, while enforcing the security requirements of all the user data.

In fact, access to de-identified patient medical data can open new avenues for revenue generation. For example, pharmaceutical companies would like to know patient demographics to study effects of their drugs and treatments plans. This revenue can be paid to the hospitals as the bulk data owners, and a fraction can be passed on to the patients whose deidentified data is made available in the Cloud. Besides the monetary transactions, the value of patients having access to their own data over their lifetime is immense. This shall be explored in the next section.

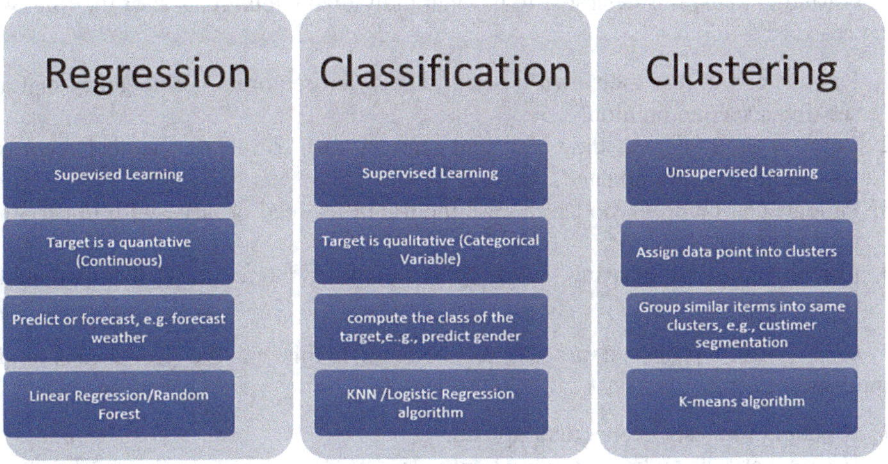

Fig. 9.1 A comparison of data Regression vs. Classification vs. Clustering [10]

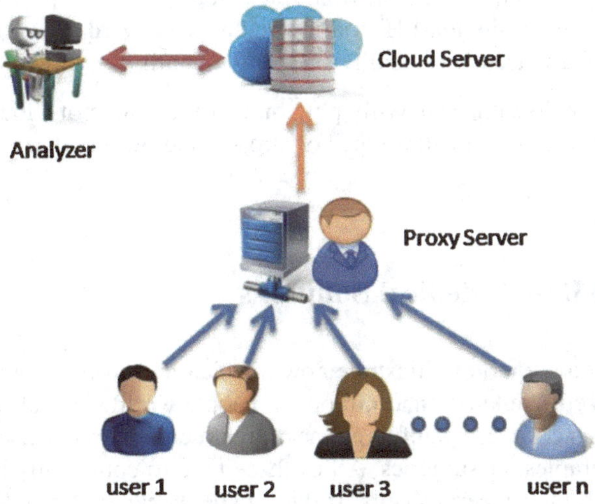

Fig. 9.2 A secure multi-party-based Cloud computing framework [11]

9.4 Need for a New Medical App

Lately, there have been several articles in the press about the need to keep track of one's own medical records [12–14]. The need for this is greater for the individuals when they are away from their home base. For example, they may need treatment while away on a business or pleasure trip. Hospitals have a lengthy process to release medical records—for good reasons such as privacy. However, in emergencies there is a risk of incorrect treatment being administered when without access to patient's medical data and history.

For individuals, it is desirable to have an individual Cloud reader for the following reasons:

1. Patients can access their own medical data when required in emergency or for seeking a second opinion.
2. Request each hospital they visit and interact with to release medical records whenever these are updated.
3. Patients can control who can access their data for read or write with in certain time bounds.
4. Enable non-PHI sharing with other commercial entities for potential monetization.

Below is the proposed process to accomplish this via the proposed Cloud application:

1. A patient requests his/her data upload.
2. Hospital checks authorization and uploads.
3. The patient's entire dataset is in the Cloud.
4. The patient decides sharing options.
5. The proposed Cloud App enables and tracks logs.
6. The patient can opt-in non-PHI sharing and gets a reward (low/no Cloud fees).
7. Hospitals, pharmaceuticals, and R&D teams can utilize non-PHI data.

An application like this can easily plug into an ecosystem of Cloud-based medical solutions currently being developed or already deployed as discussed in the next section.

9.5 Cloud-Based Medical Solutions

Let us consider a scenario to illustrate how a typical Cloud-based medical solution may be organized. We know that millions of people worldwide suffer from cancer. Once a diagnosis is done, a painful regime of treatment follows. This includes radiation, chemotherapies, or surgeries. All of these tend to collaterally impact healthy cells along with the cancerous cells. If these patients can be treated as individuals based on their specific genome sequencing, then a precision treatment can be quickly planned and the course of disease can possibly be controlled more effectively.

With the above problem in view, Intel launched a Collaborative Cancer Cloud (CCC) in 2015 (Fig. 9.3). As shown, it enabled institutions to securely share their patients' genomic, imaging and clinical data for potentially lifesaving discoveries [15]. Such sharing of data from sites all around the world could be analyzed in a distributed system, while preserving the privacy and security of the patients. The Collaborative Cancer Cloud was a precision medicine analytics platform that allowed institutions to securely share patient genomics, imaging and clinical data for potentially lifesaving discoveries. It enabled large amounts of data from sites all

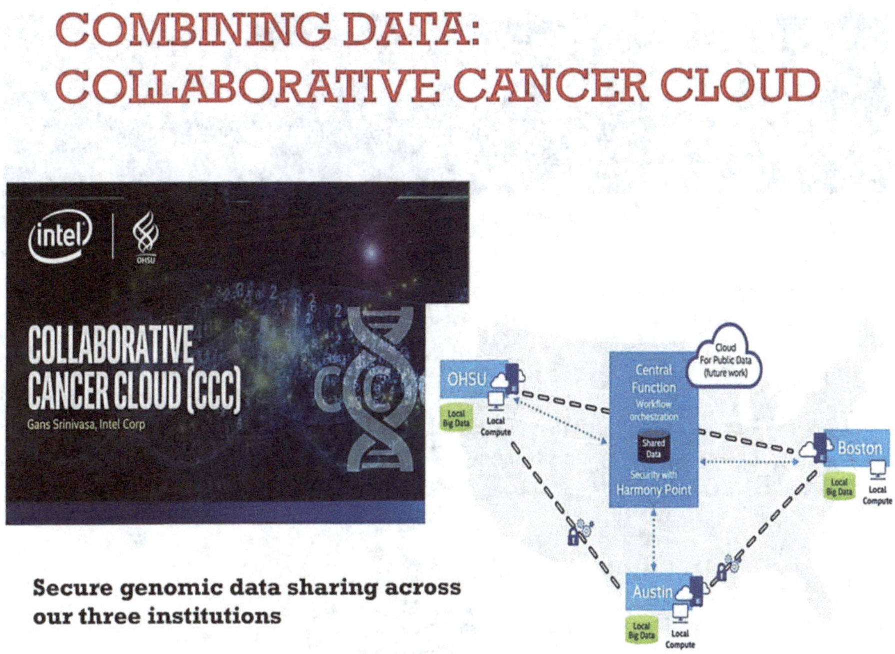

Fig. 9.3 Intel's Collaborative Cancer Cloud [15]

around the world to be analyzed in a distributed way, while preserving the privacy and security of each patient's data at each site. In exploring solutions beyond cancer, the real problem lies in the fact that we need to identify subjects for clinical trials for a robust ML solution. That will help in predicting responses of patients to proposed treatments and pick relevant biomarkers and extract new insights [16].

Hence, a new technology has emerged to train ML models at scale across multiple medical institutions without moving the data between them. It is called Federated Learning (FL) [16], an instance of which is depicted in Fig. 9.4. FL enables data to stay local while algorithms travel across the participating institutions. Note that FL helps in training a deep learning algorithm preserving privacy and security of the patient data.

For example, Intel and Univ. of Pennsylvania have a collaboration involving 29 international medical centers to train models to recognize brain tumors [17]. The long-term goal of this project is to provide diverse datasets for ML that no single institution could have provided alone. Such an approach has resulted in a new class of solutions that can identify brain tumors from a greatly expanded version of the International Brain Tumor Segmentation (BraTS) challenge dataset [18]. BraTS has been focusing on the evaluation of the state-of-the-art methods for segmentation of brain tumors in multimodal magnetic resonance imaging (MRI) scans. Adding FL [16] datasets to it will enhance the quality of results and may benefit patients around the world.

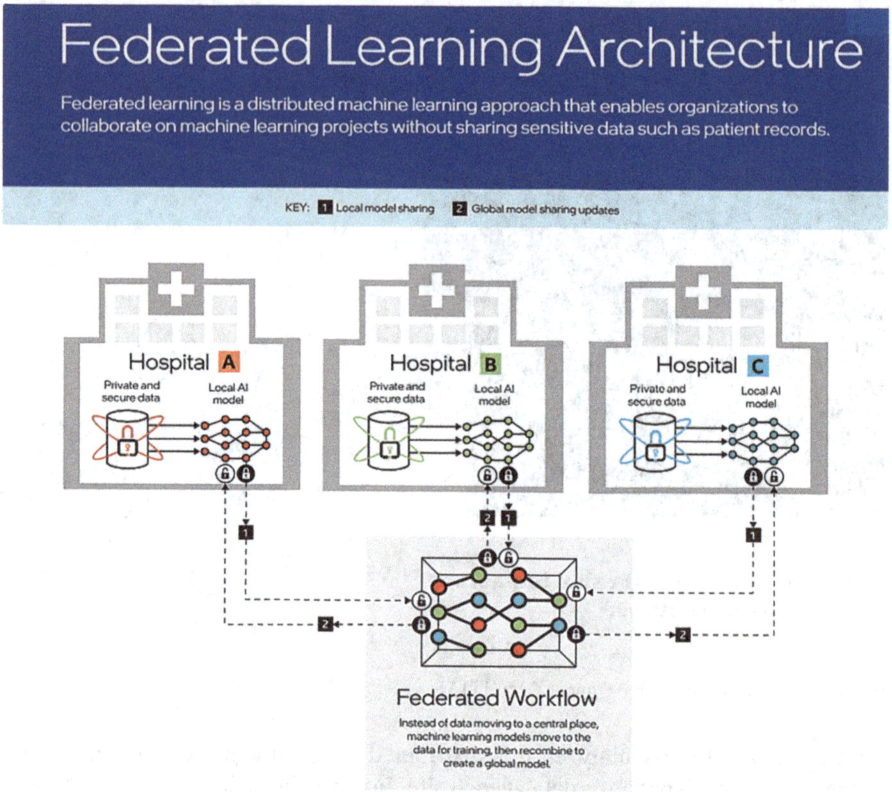

Fig. 9.4 A federated learning solution for ML and healthcare [16]

9.6 Regulations for Medical Mobile Applications

The FDA has acknowledged that mobile applications can help people to manage their own health and wellness, promote healthy living, and gain access to useful information whenever and wherever needed [19].

Mobile Medical Applications (MMA) are defined as medical devices that are both mobile apps (i.e., running on a smartphone or other mobile communication devices) and incorporate functionality, which meets the definition of a medical device by FDA. MMA is treated as an accessory to a regulated medical device. This is because sometimes it transforms even a mobile platform into a regulated medical device. FDA does not expect manufacturers to submit premarket reviews, register or list their software with the FDA when SW functions meet the regulatory requirements. This includes device software functions that help patients/users to understand and self-manage non-clinical aspects of their disease or condition.

Examples of the above include a mobile application that collects blood pressure or blood sugar levels from some FDA approved measurement device and display these on a smartphone or upload the same to a Website. However, if any data processing is done to generate an alert for an abnormal value, then a prior approval from FDA is needed. Appendix 2 has a description of the classification of medical devices used by the FDA.

This implies that if the proposed Cloud application is collecting data from various health providers and storing it, then this would be ok without FDA approval. However, if AI/ML techniques are added to interpret this data and suggest potential issues or treatment options, then the FDA approval will be required. FDA approval will also be required if a suggested treatment option ensues.

9.7 Potential Solution Setup

In the light of requirements for a new medical application and the constraints stated above, a possible solution setup to move patient data from various hospital databases after removing PHI to store it in a Public Cloud. This alleviates privacy or security concerns. Data de-identification is a difficult task because it involves removing PII (Personally Identifiable Information) from every patient data file. The difficulty arises from the variations in standard formats and medical images. It requires good OCR (Optical Character Recognition) software as a patient's name is sometimes written on X-ray and MRI screenshots. Such areas are erased, but it must be done carefully so as not to obscure the resulting medical images. In contrast, it may be trivial to add PHI back to a report, as this needs to be done only in the erased areas on each record, or sometimes only once per medical report on the first page. Figure 9.5 shows the paths of requesting (1a and 1b), de-identifying (2), storing and then subsequently accessing (3) medical records using a patient centric medical application.

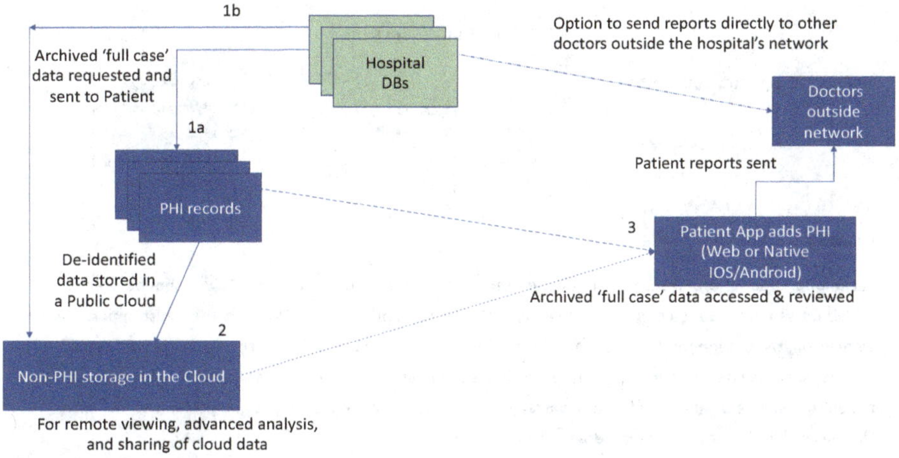

Fig. 9.5 Data flow from hospitals to Cloud and then to a patient application

The process of data de-identification begins with initially clear text data. We will describe the workflow used for de-identification of Personal and Medical examination data. Typically, this includes patient data, reports, and images. The algorithm for de-identification searches for a patient's name, date of birth, address, phone number, etc. (collectively known as PII or Personally Identifiable Information) in each document. The algorithm may choose to replace PII with another name and/or a fictitious date, or covering this information with a Whitebox, or simply replace relevant pixels with the same color as the background of an image. This action should be done carefully so that medically relevant data is not hidden or altered. It is important to ensure that the resulting images and data remain useful for the researchers.

Two examples of de-identified data are shown in Figs. 9.6 and 9.7. First is a patient data report in which a whole block of PII is covered by a Whitebox to effectively redact PHI. This protects a patient's identity, even if this report is freely circulated on the Internet. Note that the name of operator is not hidden as it is not considered PHI under HIPPA. Yet another example is that of a medical screenshot from a transcranial doppler device, showing the blood flow in the Right Middle Cerebral Artery (RMCA) for the same patient, whose patient data report was previously de-identified. In this case, pixels of the screenshot containing name and date of birth were simply colored black, same as the background color. This is done artfully so as not to alter medically relevant information in the image and resulting data can be used by the researchers.

While it is important to remove a patient's identifiable information, it is also important to replace it with a uniquely hashed token. This would enable future data and results for the same patient to be co-related, so doctors can observe progression of a disease and effectiveness of the treatment. In fact, one may even use blockchain to maintain the temporal character of patient data including the outcome of the treatment.

TCD Bubble Study Exam Report

subject id: SpontaneousEmboli

(redacted protected health information)

Exam Type:	**TCD Bubble Study**
Exam Start:	**06/01/2022 01:48:54 PM**
Exam Recorded Time:	**00:15:18**
Exam End:	**06/01/2022 02:04:15 PM**
Operator:	**Rebecca Angrisani**

Indications: CVA/TIA

Technique/Description: Transcranial Doppler exam was performed using the NovaGuide robotic system with 2-MHz ultrasound probes along with color-power M-mode and pulsed-wave Doppler. The transtemporal windows were insonated to monitor the intracranial blood flow for microbubble emboli during bacteriostatic saline injections. These injections were performed during rest and an adequate valsalva strain for 10 seconds with the patient in a supine position. Technically challenging exam due to patient paralysis and position during exam. Unable to identify left temporal window signal.

Fig. 9.6 De-identified pdf report of patient

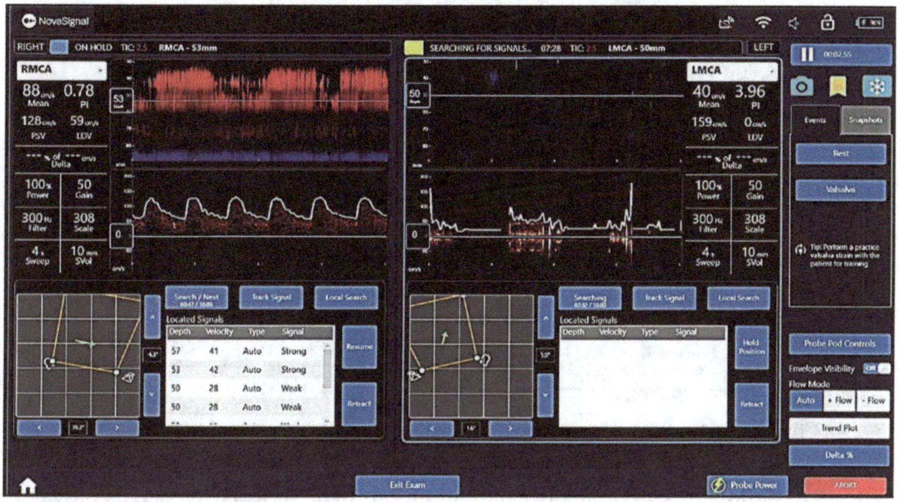

Fig. 9.7 PHI has been removed by changing pixels to background color

Once the data is stored in the Cloud, one or more of the following tasks can be performed on it:

1. **Detection:** Enabling access to external medical experts opens new possibility for early detection of disease and trends.
2. **Characterization:** A patient's medical data can be accurately characterized by.
   (a) Comparing it with the same patient's previous results from other hospitals to look for deviations of significance.
   (b) Comparing a patient's results with others for differences and detection of abnormalities.
3. **Future Predictions:** By noticing progression of a disease in other similar patients, a doctor may be able to predict the course of treatment for the same disease or manifestation of similar symptoms in each patient.

Some of this requires use of Classifiers. The design details of such Classifiers are beyond the scope of this book but are explained well elsewhere [9, 10]. However, we show two examples of managing this in Fig. 9.8. Federated classifier achieves the purpose of collaborative learning from a large amount of data that belongs to different parties without sharing the raw data among the data owners. It can utilize the Cloud computing capabilities of multiple learning agents to create different models for different patients with similar symptoms for a disease, to be shared with pharmaceutical companies and hospitals to build a global model for the progression of that particular disease.

The second classifier shown above can be used for healthy patients in a distinctly identifiable demography, wherein the same type of medical data such as cholesterol values can be sent to one compute node, and blood sugar data to another. This can

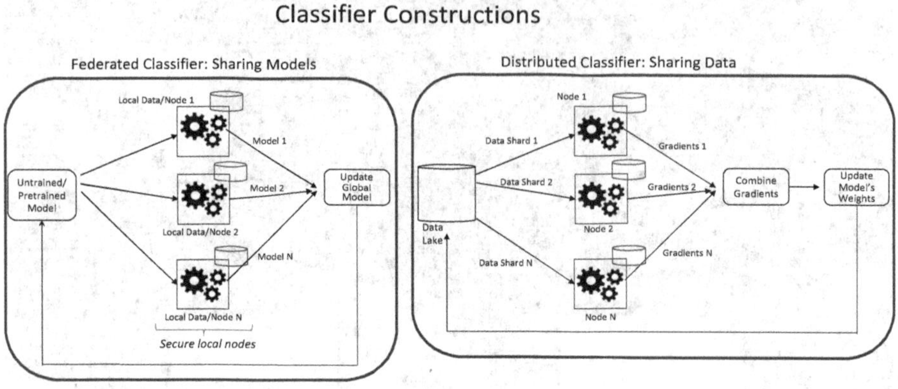

Fig. 9.8 Two possible classifiers using patients' data in the Cloud

be used to build a model of healthy individuals showing normative ranges prevalent in that demography. This helps to compare health across regions and unearth why many people in one area live longer or healthier than others. This may be due to a regional diet or genetic pool. For example, during the 1950s, when it was noted [20] that heart disease was not as common in the Mediterranean countries as it was in the USA. Since then, numerous studies have confirmed that the Mediterranean diet helps to prevent heart disease and strokes.

9.8 App Prototyping Options

The proposed Cloud application can be developed either as a Native App or as a Progressive Web App (PWA). A Native App is written to run on the operating system of a mobile device and can be downloaded from one of the many available App stores. A couple of examples are Apple's IOS and App store, or Android and Google Play Store. In contrast, a Progressive Web App (PWA) is written to run inside a Web browser, such as Safari, Mozilla, or Chrome. It is called progressive because App features can be added over time to make it look and behave like a native application.

Native Apps on a mobile platform use the programming languages of choice for that specific platform, such as Swift or Objective-C for IOS, and Java for Android. Whereas PWA usually uses a popular Web development language such as HTML, CSS, JavaScript, etc. PWA is useful when the application development team does not need to maintain multiple versions for IOS and Android devices. In addition to using a single code base, PWA is faster to build and update as it uses browser's development tools to interact with the users and bypasses mobile App store for distribution. PWA users only need a browser and URL (Universal Resource Locator) to start with.

PWA is considered more secure, as these apps always run under HTTPS protocol. This enables encrypted exchanges between a client and the server. This capability is useful for data transmission of personal health or financial information. It may be pointed out that users trust Native Apps more as these undergo additional security testing by the App store providers. In addition, App developers can build more security measures for Native Apps by adding features such as multi-factor authentication.

PWA, on the other hand, loads faster than a native app as it is just a URL and does not take much space on the device. There is no need to update an app's link as it always points to the most recent version on the server side. However, PWA may have a higher latency and more battery consumption than the Native Apps. A Native App can access device hardware to do computations and deliver a better user experience. Native code is usually faster and the resulting App is better integrated with the mobile operating systems for the look-n-feel experience.

Overall the PWAs have the following limitations:

1. ***Notifications:*** Native apps use push notifications, even to a locked screen. In contrast, PWA may have limited notification capability unless the browser has a provision for it.
2. ***Geofencing:*** Geofencing refers to virtual boundaries related to geographical areas, such as a building, city, state, or a nation. Native App developers can use these with push notifications to trigger actions such as no access to PHI outside hospitals. Currently, this is not available in PWA.
3. ***Payments:*** For mobile payments, PWA has limited access to local device hardware such as NFC (Near Field Communications) for digital payments, while Native Apps do better.
4. ***Adjacent App:*** Native Apps can link to other apps on mobile devices to check or edit calendar, make calls, send messages or mails, etc. Native apps can also use mobile device sensors to detect proximity, sound, or light conditions. Such a functionality is not available in the PWAs due to the system firewalls or other in-built browser limitations.
5. ***Offline:*** PWA works when the browser is connected to the Internet. Native apps can work well in offline mode using stored content on mobile devices. This may be important for the users always on the move and with limited Internet access in certain environments such as during a flight or in the hospital basement, etc.

In spite of these limitations, PWA is preferred in the following situations:

(a) When one needs to create a rapid prototype.
(b) When one needs to demonstrate potential of the device or some new functionality to external stakeholders such as investors and potential users.

9.9 Regulations for IoT-Based Devices

Aside from the FDA's regulations to medical device safety, efficacy and data security, IoT-based devices can have additional regulatory requirements. This stems from the fact that IoT devices are electrically powered products and often use the radio frequency spectrum to communicate. For an IoT device to be sold in the USA, it must meet compliance requirements from the Federal Communications Commission (FCC) and should meet applicable domain and product safety requirements based on industry standards. Similar requirements and standards are applicable if the device is to be sold in other countries.

The FCC was established [21] to regulate interstate and international communications by radio, television, wire, satellite, and cable. FCC encourages the highest and best use of the electromagnetic spectrum domestically and internationally. Regulations addressing the use of the spectrum and public safety are published in Title 47 of the Code of Federal Regulations (CFR) [22]. The applicable rules and regulations are an additional watermark that every IoT company must satisfy before being allowed to sell in the USA.

There are many documents available to aid in FCC compliance [23, 24]. These include industry-specific guides for small businesses from the FCC itself [25]. At a high level, the steps include:

1. Identify the allowed frequency ranges in the electromagnetic spectrum and whether a license is required.
2. Adhere to the technical requirements, such as maximum emission strength, for the given frequency range.
3. Label product according to FCC rules.
4. Test product to ensure compliance.
5. Maintain adequate records of design and test.

Let's illustrate this with a few examples. Internet connected wildlife tracking devices are a valuable tool for environmental researchers and conservationists. These have been used to track salmon in the US Pacific Northwest as part of habitat restoration efforts [26]. Similarly, there are recommended guidelines to combat the poaching of tigers in the wild [27]. The FCC specifies the frequency bands of 40.66–40.70 MHz and 216–220 MHz as among the allowable frequencies for tracking animal wildlife [28]. The FCC requires users to obtain a license to operate in these bands. A license is a government issued permit that may need to be presented for lawful radio communications. Additionally, standard Radio Frequency Identification (RFID) bands such as 902–928 MHz could be used without a need for a license (47 CFR Parts 15.1 and 15.247, [29]). India's Department of Telecommunication allows for the use of animal tracking with the Ultra High Frequency (UHF) band of 865–867 MHz for RFID, also without the need for a license [30]. It is important for IoT product development teams to understand and conform to the relevant communication requirements.

Here is yet another example. Commercial space launch companies may wish to install Internet-connected ground-to-launch-vehicle telemetry devices. These can monitor the trajectory of a rocket at various stations across the Earth, or enable space enthusiasts to track the launch. As stated in 47 CFR Part 26 [31], space launch service devices are allowed to use frequencies in the range of 2025–2110 MHz and 2200–2290 MHz for communication. However, the team must be aware that only the 2200–2290 MHz band is available specifically for telemetry and tracking operations per Section 26.3 of Part 26. Since the frequencies allowed for space launch services are also shared with other industries such as fixed mobile and satellite communication, the telemetry devices should meet the power transmission limits set forth by the FCC in Section 26.303. Additionally, the telemetry tracking devices may also use any of the spectrum bands dedicated to Wi-Fi (2400, 5000, 6000 MHz) for Internet communication.

In the medical space, a personal medical health IoT device that monitors physiological signals across the body (chest, limbs, head) with different sensors may communicate in the 401–457 MHz band or the 2360–2400 MHz band [32]. The FCC specifies the 2360–2400 MHz band is a secondary use for medical body area networks (MBANs). This means that other primary users of this part of the spectrum including aeronautics, amateur radio, and radio astronomy should not be interfered with by the medical IoT device. This is achieved with a lower emission limit allowed for MBANs as set by the FCC. An IoT device operating in the MBAN secondary band will need to adhere to these limits for the 2360–2400 MHz band but may also any of the spectrum bands dedicated to Wi-Fi (2400, 5000, 6000 MHz) for Internet communication. Transmitters used in the IoT device should be labeled per FCC requirements.

While this may seem overwhelming for a IoT product development team, wireless modules that are compliant out-of-the-box are available and can be integrated into IoT devices. Figure 9.9 shows an image of ESP8266-based 2.4 GHz Wi-Fi module. These modules are affordable, meet the FCC's technical and labeling requirements, and can include integrated PCB antennae.

Regarding electronic product safety for IoT devices, obtaining Underwriters Laboratories (UL) certification is prudent. While adherence to UL standards is voluntary, many product distributors will not sell a product unless it has a UL mark (or equivalent), and many businesses will not install a product without such certification [33]. UL standards specify product design rules, construction techniques, and

Fig. 9.9 FCC Compliant Wi-Fi module

performance guidelines that minimize risk to various forms of hazards such as electrical shock or fire. Safety standards are maintained by other organizations such as the American National Standards Institute (ANSI), the American Society of Mechanical Engineers (ASME), and the American Society for Testing and Materials (ASTM) and can be used instead of UL standards. Many of these standards are "harmonized" with international standards—meaning compliance with a standard can suffice for multiple countries or regions. A similar process to meet FDA or FCC requirements applies to safety standards. Relevant sub-clauses of the standards should be identified, tested against and documented.

A project team must consider the time and cost of FCC and UL compliance. Each round of testing against regulatory standards can cost between $5k and $20k and multiple rounds of testing may be necessary during product development. Formal test reports can take up to 4 weeks to be completed by test laboratories after testing has been completed, which can take several weeks on its own.

9.10 Summary

Healthcare is an important frontier for humankind to win over existing and new diseases. This requires collective and massive effort to engage in data analysis. Enormous possibilities exist using Cloud-based data repositories and ML tools. In the near future, AI techniques could describe and predict the course of diseases, along with a recommendation for treatment. Obviously, these can be presented to the doctors based on the training with previously labeled data.

We reckon that it is of vital importance for a patient to have full access to his own health records across different hospitals during his lifetime. Here we have outlined a Cloud-based application that enables such access. This would be most useful for seeking emergency care away from the home base or for seeking a second medical opinion.

Appendices

Appendix 1: Regulation of Clouds in India

It has been said that data is the new oil of twenty-first century and thus should be protected as a valuable resource of information [8]. Cloud data operations spread over multiple geographies with much larger attack surface. Privacy is a key issue related to the adoption of Cloud computing. If a private datacenter owned by an enterprise is attacked, information pertaining to only that enterprise is compromised. Whereas a public Cloud may contain data of many different entities, so the scope of damage is much larger. Hence, data security issues pertaining to Confidentiality, Integrity, and Availability (CIA) need to be regulated and aligned to the geographical laws. Western countries already have laws on their books, such as HIPPA in the USA, and GDPR in European Union, to protect its citizens' PHI (Patient Health Information) and PII (Personally Identifiable Information), respectively. Among the non-western nations, India has a large IT industry that supports Cloud-based operations. India also has a fairly robust information technology act, promulgated and revised periodically since the year 2000.

Quite recently, Supreme Court of India stated that the right to privacy is a fundamental right as it is derived from the rights that are enshrined in the constitution of India [34]. It emerges from the right to life and personal liberty.

Under the Indian Telegraph Act, 1885 telegraph includes "any appliance, instrument, material or apparatus used or capable of use for transmission or reception of signs, signals, writing, images, and sounds or intelligence of any nature by wire, visual or other electromagnetic emissions, Radio waves or Hertzian waves, galvanic, electric or magnetic means." The same could be used to construct the definition of Cloud as the Cloud is a means by which a person is able to send and receive data which is operated using communication networks [35].

While in this Act, privacy is recognized as a right, but the government has the power [36] to intercept the communication in matters relating to national security [37]. The Information Technology Act, 2000 explicitly provides the penalties for breaches of data and privacy of private individuals using computers and technology. This Act focuses on cybercrime and e-commerce in general but data protection and data privacy are covered under it. Sections 43, 65, 66, and 72 specifically provide the rules for data protection.

India's Information Technology Act, 2000 provides conditions under which an intermediary could be held liable. An intermediary would not be held liable, if it could be proved that it has not originated the information targeting the audience or has altered the information, and it is merely transmitting the information in as is mode only. Further, if there is any violation of data and if intermediary has been informed then the information has to be taken down. If the intermediary fails to take down the information or has abetted, aided, or conspired for the commission of an unlawful act the intermediary could be held liable.

In Cloud, there is also an issue of jurisdiction as multiple countries may be involved, and their laws may be in conflict. According to the civil procedure code, a case can be initiated in two places, at the residential place of the defendant, or where the action arises. This does not define where cases of cybercrimes may be dealt with, as there is no physical place of crime. In India, if an act is committed in contravention of Information Technology Act, 2000 by a person who is not a resident of India but involves a computer, computer system, computer network located in India, then the Indian act is applicable on that person or entity [38].

Our recommendation for US companies with employees in India is to incorporate in India as well to protect its intellectual property. Note that 66% of all data breaches are caused by insider threats [8].

Appendix 2: FDA Classes of Medical Devices

The Food and Drug Administration (FDA) classifies all medical devices into three categories based on their potential risk to the patients, as follows:

1. ***Class I:*** These devices are low-risk devices. Examples include bandages, hand-held surgical instruments, and nonelectric wheelchairs.
2. ***Class II:*** These devices are intermediate-risk devices, such as Ultrasound or X-ray machines, and electrical wheel chairs.
3. ***Class III:*** These devices are high-risk devices such as automation used in surgery. Clearly, such devices are very important to health or sustaining life requiring Pre-Market Approval (PMA). The examples include remote controlled surgical robots.

Points to Ponder

1. What's the role of regulations in the medical industry?
2. In which cases software is treated as a separate medical device, and in which cases it is treated as adjunct to a medical device?
3. Why FDA has reservations in using AI/ML-based techniques for medical diagnosis?
4. What is the medical value of data de-identification?
5. Is it sufficient to remove a patient's information for broad data sharing?
6. Should a patient be given access to his own medical data?
7. Why do IoT device radio communications need to be regulated and by which agencies?

Answers

1. What's the role of regulations in the medical industry?

   (a) The role of medical regulation is to protect the interest of medical service consumers and patients.

2. In which cases software is treated as a separate medical device, and in which cases it is treated as adjunct to a medical device?

   (a) If a software application is used to collect data from another medical device, and display it along with historical values, then the software is not treated as a separate medical device. However, if the software application is used to identify abnormal values, or give interpretations of the data then it is treated as a separate medical device and FDA approval is required.

3. Why FDA has reservations in using AI/ML-based techniques for medical diagnosis?

   (a) As such the FDA is not opposed to the use of AI/ML techniques being used for medical diagnosis but does require presence of a human element in final decision-making. This is no different from the role of software in self-driving cars. In case of an injury or accident, the question of liability arises and needs to be addressed. Invariably software providers or car manufacturers may not be willing to accept that kind of liability. It is for this reason the manufacturers such as Tesla currently make it mandatory for human drivers to be always behind the wheel even in the auto-pilot mode. For the same reason, the FDA requires a medical doctor to certify reports and results for the final diagnosis. However, doctors can use the software to assist and speed up their decision-making process. Consider the example of examining X-ray or MRI results of patients with bone injuries. Orthopedic analysis support software can highlight current as well as potential locations with displacement, gaps or cracks, etc. However, in the end a doctor must examine and agree with the suggestions before treatment can begin.

4. What is the medical value of data de-identification?
    (a) Medical science advances by building upon knowledge of how patients react to the treatment and drawing generalized inferences for applying this knowledge to the other patients with similar symptoms. However, due to human biases, sometimes a patient may feel discrimination in their careers or society, if the news of their diseases is spread beyond a strict need to know basis. Another example is a future employer not wanting to hire a heart patient, thinking it will increase future healthcare costs for the company. Similarly, health information of political leaders is often a highly guarded secret. For instance, Pakistan's founder Mr. Jinnah's deteriorating health condition was kept highly secret. Even today some people argue that if this information was known publicly, then partition of India may not have happened.[1] Thus, medical data needs to be de-identified before being shared across the medical community, as not sharing any data is detrimental to the progress of medical sciences.
5. Is it sufficient to remove a patient's information for broad data sharing?
    (a) While it is important to remove a patient's identifiable information, it is also important to replace it with a uniquely hashed token. This helps in co-relating data and results for the patient, so doctors can observe progression of a disease and efficacy of a treatment.
6. Should a patient be given access to his own medical data?
    (a) The current practice is to share medical data with the patient on a need-to-know basis. Also, since a patient may not be medically savvy, there is a tendency of doctors to up-level or obfuscate the details. This becomes a hindrance if a patient needs to seek healthcare away from the home base. Hence, there is a need to give full access to a patient for his own medical records. This is fully supported by HIPPA.

[1] https://www.amazon.com/Freedom-at-Midnight-Larry-Collins/dp/0006388515

7. Why do IoT device radio communications need to be regulated and by which agencies?
   (a) IoT device radio communications need to be regulated in order to promote the highest and best use of the electromagnetic spectrum within and between countries. Some of the electromagnetic spectrum is reserved for government use in support of public safety, air traffic control, or national defense. Other restricted uses may be for public television or radio. Regulatory agencies allow a broad range of dedicated or shared spectrum for commercial devices including IoT, ensuring that critical public services are not disturbed by IoT device radio communications.
   (b) The FCC, UL, ANSI, ASME, and ASTM are some examples of the agencies that provide mandatory or voluntary regulations of radio communications. The FCC regulates the use of the electromagnetic spectrum in the USA. It specifies which frequency bands can be used and how they can be used by products given a particular application. IoT devices must comply with FCC regulations to be sold in the USA. UL, ANSI, ASME, and ASTM provide standards for product safety that are often considered required for sale.

References

1. https://www.fda.gov/
2. https://www.consumerreports.org/cro/magazine/2012/04/cr-investigates-dangerous-medical-devices/index.htm
3. https://www.healthcareitnews.com/news/emea/unsafe-medical-devices-allowed-global-markets-harm-patients-investigation-reveals
4. https://oag.ca.gov/privacy/ccpa
5. https://gdpr-info.eu/
6. https://www.hhs.gov/hipaa/index.html
7. https://www.businessinsider.com/theranos-founder-ceo-elizabeth-holmes-life-story-bio-2018-4
8. Sehgal, N. K., Bhatt, P. C. P., & Acken, J. M. (2020). *Cloud computing with security and scalability*. Springer. https://link.springer.com/book/10.1007/978-3-031-07242-0
9. Dunham, M. H. (2002). *Data mining: Introductory and advanced topics*. Prentice Hall PTR.
10. Gupta, P., & Sehgal, N. K. *Introduction to machine learning in the cloud with python*. Springer. https://link.springer.com/book/10.1007/978-3-030-71270-9
11. Pussewalage, H. S. G., Ranaweera, P. S., Oleshchuk, V. A., & Balapuwaduge, I. A. M. (2013). *Secure multi-party based Cloud Computing framework for statistical data analysis of encrypted data*. ICIN 2016, At Paris. http://dl.ifip.org/db/conf/icin/icin2016/1570221695.pdf
12. https://www.cnet.com/health/medical/how-to-keep-track-of-your-medical-records/
13. https://www.verywellhealth.com/how-to-get-copies-of-your-medical-records-2615505
14. https://www.drugwatch.com/health/how-to-obtain-medical-records/
15. https://itpeernetwork.intel.com/intel-ohsu-announce-collaborative-cancer-Cloud-at-intel-developers-forum/
16. https://owkin.com/federated-learning/
17. https://www.hpcwire.com/2020/05/11/intel-upenn-launch-massive-multi-center-ai-effort-to-refine-brain-cancer-models/
18. http://braintumorsegmentation.org/

19. https://www.fda.gov/medical-devices/digital-health-center-excellence/device-software-functions-including-mobile-medical-applications
20. https://www.mayoclinic.org/healthy-lifestyle/nutrition-and-healthy-eating/in-depth/mediterranean-diet/art-20047801
21. https://www.fcc.gov/
22. https://www.ecfr.gov/current/title-47
23. https://compliancetesting.com/fcc-certification-requirements/
24. https://macronetservices.com/fcc-19-72a1-compliance-checklist-ensure-your-business-is-compliant/
25. https://www.fcc.gov/communications-business-opportunities/compliance-guides-small-businesses
26. https://www.cascadepbs.org/environment/2021/10/wa-fish-researchers-use-tiny-sensors-and-other-tech-save-salmon
27. https://conservewildcats.org/2024/02/29/iot-and-lora/
28. https://www.ecfr.gov/current/title-47/chapter-I/subchapter-D/part-90/subpart-J/section-90.248
29. https://www.ecfr.gov/current/title-47/chapter-I/subchapter-A/part-15
30. https://www.gma.trade/news/india-introduces-exemption-from-license-for-some-rfid-devices
31. https://www.ecfr.gov/current/title-47/chapter-I/subchapter-B/part-26
32. https://www.fcc.gov/document/medical-body-area-networks-first-report-and-order
33. https://certification-experts.com/electronic-product-united-states/
34. Consultation Paper on Cloud Computing. http://www.trai.gov.in/sites/default/files/Cloud_Computing_Consultation_paper_10_june_2019.pdf
35. Section 5 of the Telegraph Act, 1885.
36. Reeta Sony A. L, Prof Sri Krishna Deva Rao, Bhukya Devi Prasad, Implications of Cloud computing for personal data protection and privacy in the era of the Cloud: An Indian perspective. https://www.researchgate.net/profile/Reeta_Sony/publication/262049549_Implications_of_cloud_computing_for_personal_data_protection_and_privacy_in_the_era_of_the_Cloud_An_Indian_perspective/links/0f31753b28664618d9000000/Implications-of-cloud-computing-for-personal-data-protection-and-privacy-in-the-era-of-the-Cloud-An-Indian-perspective.pdf
37. Paragraph A, Chapter 2, Pg. 15 of The report issued by the Committee of Experts under the Chairmanship of Justice B.N. Srikrishna on Personal Data Protection.
38. https://www.idwatchdog.com/insider-threats-and-data-breaches

Chapter 10
Hybrid Cloud-Based Applications and Security

10.1 Introduction

A hybrid Cloud refers to the combination of an on-premises datacenter (aka private Cloud) with a public Cloud [1]. It allows data and applications to be shared between the private and public Clouds. Hybrid Cloud is the fastest-growing segment as it enables enterprise ITs and even SMB (small and medium business) owners to operate their onsite computing facilities, and leverage advantages of a public Cloud. For example, some hospitals may not wish to put their patients' Private Health Information (PHI) in a public Cloud. Note that Personally Identifiable Information (PII) is a superset of PHI. So, they remove PII and PHI, replace it with a unique token, and put de-identified data in the public Cloud [2]. PHI data is maintained in the private Cloud, and any diagnosis done using public Cloud data is then matched back to the appropriate patients using unique tokens.

In general, a hybrid Cloud may offer the following benefits:

1. **Business Continuity:** In the case of an external Internet disruption and limited access to outside Cloud resource, users can use the private Cloud. Similarly, during sudden demand surges or for internal infrastructure upgrades, users can migrate their jobs to a public Cloud. This mode is also called Cloud bursting [3].
2. **Cost Optimization:** Hybrid Clouds help to create a single, flexible, and cost-optimal IT infrastructure [4].
3. **Regulatory Requirements:** Hybrid Clouds help enterprises to comply with regulatory requirements, such as HIPPA laws [5], and data sovereignty requirements such as GDPR [6].
4. **Technology Investments:** Many organizations may already have invested in on-premises infrastructure before migrating to public Clouds. A hybrid solution allows them to take full advantage of previous investments, many of which may have already been fully depreciated.

P. C. P. Bhatt et al., *Project Management in Cloud and IoT Applications*,
https://doi.org/10.1007/978-3-031-99427-2_10

5. **Low Latency:** Access times to public Clouds depend on the Internet bandwidth and distances to the nearest datacenters. A hybrid Cloud can help to mitigate this delay with relatively lower latency on-premises resources for mission critical applications.

Some enterprises use hybrid Clouds in lieu of a multi-Cloud solution using different public Clouds, thereby simplifying their operations and accounting costs. As we will see in the next section, this further alleviates the security and performance concerns in Cloud-based applications.

10.2 Limitations of Cloud-Based Applications

Like any technology, there are pros and cons to Cloud computing [7]. One of the major sources of issues is reliance on the public nodes in Internet connections from users to a Cloud Service Provider (CSP). On-premises computing may be using hardwired connections to access data on servers or storage devices. However, with Cloud computing, a poor Internet connection can limit access to the mission critical data or applications.

Even top CSPs can experience downtime due to a natural disaster or slower performance caused by an unforeseen traffic and technical issues that might impact connectivity. Other disadvantages of Cloud computing include:

- Real or perceived security risks due to remote data privacy and online threats.
- Risks of vendor lock-in, limiting migration to other CSPs in future.
- Unforeseen costs and unexpected expenses, if a CSP raises prices.

All these and a host of other factors lead many organizations to opt for hybrid applications. These applications use hybrid Cloud infrastructure, with the key difference that a part of the application's code and data may reside on a user's end-point device such as a cell phone or laptop. A common usage model is mail applications. Users may want to download and access their message to read during a flight while in the airplane mode with no connection to the remote Cloud. They may also respond to, or send new, messages during the flight. Upon landing, once Internet connection is available, users expect their devices to automatically synchronize with the mail servers in a seamless manner. This is only possible with a proper architecture of hybrid applications described in the next section.

10.3 Architecture of Hybrid Applications

Hybrid applications combine native device and web technologies to offer solutions that can access local hardware features, while still using web-based resources in the background. For any mobile application, before choosing an architecture, the following questions need to be answered [8]:

- Target audience—who will be using the app?
- Target platforms—which devices will be targeted?
- App features and functionality—what features will be in the app? What hardware features (in case of a smart phone: microphone, camera, GPS, etc.) will be used?
- Timeline and budget—how fast one needs an app and how much to budget for it?
- Developer skills—what languages and platforms developers already know?

Following is a classification of mobile and IOT applications:

1. Native applications: Native apps are binary executable applications that are stored on and run locally on a device. Native apps work just like any other built-in app and have full access to the device APIs and features. Native apps are usually obtained via an app store (i.e., the iOS App Store or Android Marketplace).
2. Web applications: Mobile web applications run inside and are (usually) accessed via visiting a URL in the web browser on a mobile device. Mobile web apps are built using HTML, JavaScript, and CSS, or a JavaScript framework that eases mobile web development. Mobile web apps can usually mimic the look and feel of a native app, though they have lesser access to device features and APIs than a native app. Web applications are constrained by running within the web browser. As the code for a mobile web app is run on a web server and not distributed via any app store, the code can be freely updated and redeployed without any submission process or approval from an app store vendor.
3. Hybrid applications: Hybrid mobile or IOT applications can use a few different architectural options, but the common thread between them is that they are largely based on web technology (i.e., HTML/CSS/JavaScript) running within a native app, mimicking the look and feel of a native app.

Hybrid apps are often wrapped local web applications, which allow the execution of native code. An example is shown in Fig. 10.1, listing metrics to measure performance [9].

As hybrid apps at their core are web applications, they can use UI toolkits to display user interfaces. However, because these UI toolkits can only imitate certain behaviors of native controls, they sometime lack the native look and feel that most users expect [10].

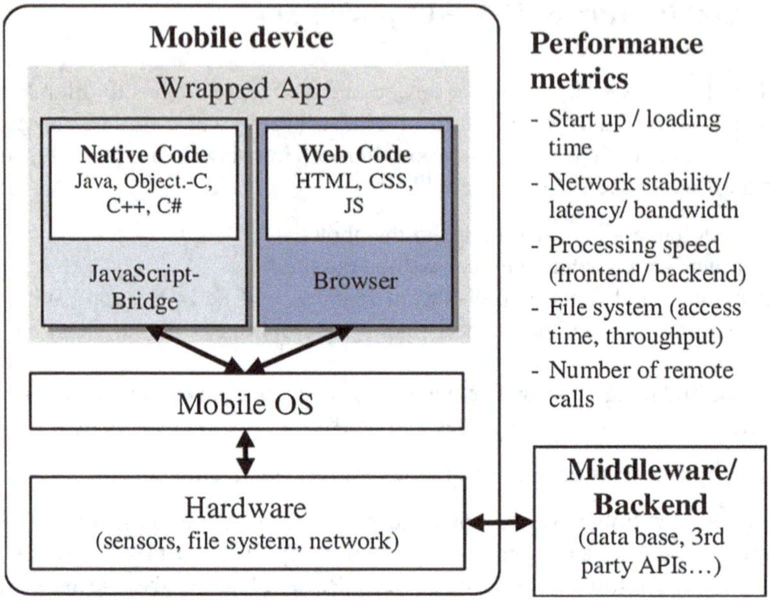

Fig. 10.1 Hybrid mobile application architecture [9]

10.4 Some Examples of Real-Life Hybrid Applications

The mobile applications industry is moving fast with new and powerful software features. Thus, it is no surprise that most highly used and high-performance hybrid applications also currently appear on mobile devices [11]. In essence, a hybrid application is a web application built using HTML5 and JavaScript, in a native app shell. This allows hybrid apps to be distributed through the app stores, while still providing access to all device features. Below is a list of some hybrid applications that provide impressive performance, near native user interfaces, and excellent functionality:

1. Gmail: Google's mail application on a mobile phone combines native and web features in a seamless manner. It allows users to download messages when connected and read/respond to them even when the phone is disconnected from the Internet. Later, the Gmail does automatic synch to send any stored messages and download new ones. The memory space on the mobile phone is protected from other applications to avoid intrusion or modification to user's mailbox. If the uploaded message is a part of an existing thread, then Gmail takes care to add the reply appropriately to the existing thread.
2. X (formerly known as Twitter): Twitter's social media application is hybrid in nature. It is quite reliable to keep the message and response threads intact in offline mode, while downloading some media on the device to play posted

videos when disconnected from the Internet. This application also shows how a vast amount of traffic can be handled by hybrid apps.

3. Instagram: A hybrid approach allows Meta's Instagram social application to support data consumption in the offline mode and process vast amounts of media.
   As we notice, all the examples given above are in the IT or social media domains. A key reason is to avoid financial risks. Imagine an application running on a bank's remote ATM location, which checks a customer's balance and allows funds to be withdrawn even during the offline mode. Funds may have already depleted by a cheque being cleared or customers withdrawing money from another ATM. In this scenario, bank may be liable for overpayment and ends up with a negative balance in customer's account upon nightly reconciliation. Hence, bank remote ATM allows funds to be withdrawn only in a connected mode to the central servers.
   However, there is a financial application that is using hybrid architecture.
4. PayPal: It uses a hybrid architecture that combines new microservices with existing monolithic systems. This allows for incremental modernization, while still taking advantage of the stability of the monolith [12]. Some of the benefits include:

   (a) Incremental Migration: for organizations with existing applications, a hybrid architecture allows for gradual migration from a monolithic system to microservices. This is accomplished by first identifying areas that will benefit from the flexibility of microservices, then new components are developed and integrated alongside the existing monolithic applications. This minimizes disruption and risks.
   (b) Leveraging Existing Investments: Monolithic applications may have been refined over time, with functionality and logic integral to a business. By integrating microservices, it avoids a complete rewrite of the application, while preserving the IP embedded in the monolithic components.
   (c) Improved Scalability: Microservices provide scalability for specific functions with an application. By isolating performance bottlenecks and implementing these as microservices, organizations can independently scale them. This approach allows for better responsiveness to fluctuating user demands and more efficient resource utilization.
   (d) Flexibility and Innovations: Microservices offer implementation options across different programming languages, frameworks, and technologies. This enables organizations to embrace modern technology and practices, while keeping the stability of legacy systems.
   However, a hybrid application's architecture also comes with some challenges, as enumerated below:
   (a) Complexity: A hybrid application's architecture introduces complexity, as the organizations strive to maintain seamless integration and communication between microservices and monoliths. A well-defined interface and protocols are needed to maintain a cohesive system.

(b) Data Consistency: As data is shared between the monolithic components and microservices, an organization needs a strategy for data management and integration. This is essential to maintain system level integrity and prevent data inconsistencies.

(c) Operational Overhead: Organizations that want to migrate from web-based or native applications to hybrid need to invest in two distinct paradigms. They will need to maintain necessary infrastructure and deployment for both microservices and monolithic components.

Next, we will visit the performance and maintenance aspects of hybrid applications.

10.5 Reorganization for the Hybrid Applications

We previously explained in Chap. 1 about the linkage between an organization's structures and its product system designs [13]. Note that sometimes both need to change for migrating a native application to hybrid Cloud. These changes may be of the following nature:

1. **Structural changes:** Since skills need to develop IoT devices and embedded devices are quite different than the skills needed to develop Cloud applications, an organization may need two distinct teams. First team can deal with the legacy applications, and a second one is needed to work on the Cloud. There needs to be clear decisions between the roles and data interfaces between the two teams. A cross-team architect is needed to make decision on the partition of responsibilities between the two teams. Ideally, this architect will be at the same level as the leaders of each team, so impartial decisions can be made. Often a committee is setup to debate technical matters and make recommendations, but the final decision and its consequences must rest with the cross-team architect.
2. **API changes:** A new set of APIs need to be clearly defined on how the Cloud part of the application will communicate with the native or embedded IoT part of the application. Often the backend code for authentication or implementation may remain the same but a different frontend and channels to access may be needed. An example is for credit card and other modes of processing financial transactions in a bank. With the introduction of Google Pay [14] or Apple Pay [15], a virtual card is created so that a user can use it for a transaction on the web without having to swipe the physical credit or debit card. Both Google Pay and Apple Pay rely on tokenization for secure transactions. It replaces sensitive card information with a randomly generated, unique sequence of numbers, called a token [16]. Using it reduces the risk of fraud, as depicted in Fig. 10.2.
3. **Product validation changes:** Clearly, both the native and embedded parts of the application need to be tested separately from the Cloud application. Then an integration testing is needed for the whole end-to-end system.

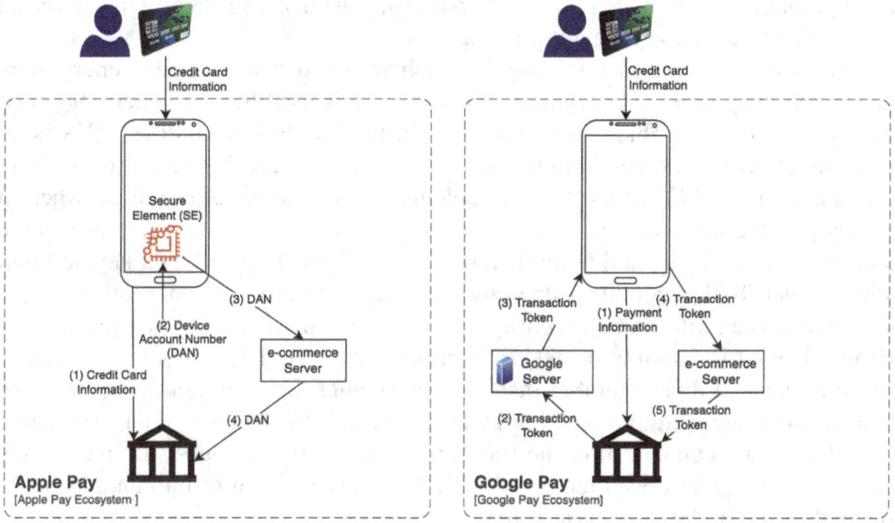

Fig. 10.2 Architectures for Apple and Google Pay [16]

4. **Release process and regulatory compliance:** It may happen that different part of the system change independently, so there should be a provision to release them separately. An example is upgrade of the IoT components, or migration from one Cloud vendor to another. In this case, a separate regulatory clearance may need to be obtained for the changed components only. Furthermore, some regulations may restrict data movement across geographies. For example, GDPR [6] requires that PHI or PII of subjects in EU need to be stored in EU datacenters only. For a medical device deployed in a European hospital, its data needs to be either de-identified before moving to a server in the USA, or the Cloud application based on the USA needs to be deployed in Europe.

10.6 Security Challenges and Solutions

Three key elements of hybrid application security are public key encryption, private key encryption, and secure hashing. We will introduce each of these and details can be found here [1]. The examples chosen (such as RSA, AES, and SHA-3) are commonly used in the current Cloud computing deployments.

A symmetric encryption requires a unique key for each pair of communicators. This increases the number of keys a person must maintain for communicating with lots of different people. For many people to communicate with each other, this method is not scalable, and key management becomes an issue. Asymmetric encryption uses a different key to encrypt than to decrypt. The name public key refers to an encryption key being made public, i.e., published openly. The RSA algorithm is an example of an asymmetric encryption scheme [17].

Symmetric encryption operation is much faster than asymmetric encryption. However, for symmetric encryption, the same key is used for both encryption and decryption, so the problem is to secretly get the key to both parties. AES is an example of symmetric encryption that is widely used in the Cloud and throughout the Internet [18]. AES encryption is implemented as a series of n rounds, where n depends on the key size. Each round consists of a byte substitution step or layer, a shift row step or layer, and a mix column step or layer followed by a key addition step or layer. To decrypt, the rounds are in the reversed order for encryption.

Hashing is an information security technique to mark a message to prevent tampering. The information to be hashed is broken into specified size blocks and combined to create a digest number that is much smaller than the input. The raw data and its digest are sent to a recipient, who can rehash the received data, compute a new digest, and compare it to the transmitted digest. If two digests do not match, then the message has been tampered with. It is very difficult to find another input that will result in the same hash value.

Information security includes the protection of information from reading, alteration, and falsification. Encryption is a key tool for protecting information from unauthorized reading, and hashing is a key tool for preventing alteration or falsification. Specifically, the symmetric encryption example of AES is used for securing large amounts of data for storage or transmission. The asymmetric example of RSA is used for exchanging small amounts of information, most commonly a method to securely exchange keys, which are required for symmetric encryption. Finally, SHA-3 is an example of secure hashing that protects information from alteration or falsification [19].

A Virtual Private Network (VPN) can be used to securely access a company's internal applications and data, including those hosted in the Cloud. A VPN creates an encrypted connection between a user's device and a server, hiding the user's IP address and encrypting their data. This prevents web traffic from being exposed to the open Internet. Thus, an enterprise can use VPN for the following services:

- Give remote employees access to internal applications and data.
- Create a single shared network between selected office locations.
- Encrypt online traffic and data as it travels between a device and the server.

Protocols, algorithms, and code for encryption can be found here [20].

10.7 Hybrid App Performance and Maintenance

Debugging a hybrid application can be quite a challenge. This is because each component is developed and tested separately. However, when integrated a hybrid application may work but with some performance issues that may need to be resolved. A common example is slow response time to user clicks. A potential root cause of performance problems is often different than the functional issues, which may be tricky to debug. Note that hybrid apps are web applications running in a native container, and inefficient code execution can lead to performance issues. These issues manifest as slower response times and laggy animations. One way to overcome this is by using high-performance frameworks, such as React Native or Flutter for complex UIs and animations [21].

It is demanding to isolate performance issues specific to each platform. To simplify the process:

- Utilize remote debugging tools for real-time app inspection and debugging.
- Implement continuous integration (CI) pipelines for automated testing and deployment on various platforms.
- Utilize emulators and simulators for quick testing across different platforms before real device testing.

To diagnose performance problems in hybrid apps, several factors need to be examined such as CPU usage, memory consumption, network latency, and rendering performance. Once measurements are made, several steps can be taken to improve the performance of hybrid apps. Foremost, it is important to examine the impact of third-party plugins and libraries, to reduce redundant operations and improve app speed. Below are some suggestions:

1. Reduce the number of plugins and libraries
2. Compress media files
3. Preload or cache content
4. Optimize application code
5. Use in-memory databases and high-speed storage for frequently accessed data
6. Continuously monitor and analyze application performance metrics
7. Be aware of noisy neighbors in a public Cloud [1]

Note that even if an application appears to work as desired at the beginning, its performance may drag over time as the number of users grows, or the size of database increases. It is often difficult to reproduce performance issues. Therefore, it is important to continuously monitor an application's performance and capture the metrics in a log file, for a later debugging session.

A modular architecture will allow partitioning of problems and solutions. This will also enable updates in a part of the system, without changing other parts to minimize perturbations if interfaces between parts are well defined and documented.

In general, maintaining hybrid applications is easier than native applications [22] because developers only need to manage one codebase that works across different hardware platforms:

- Updates and bug fixes: Updates and bug fixes can be applied across all platforms at once, ensuring consistency and reducing the time and effort required for maintenance.
- Cost savings: The streamlined maintenance process can result in higher cost savings over the app's life cycle.
- Faster time to market: Hybrid apps can be launched up to three times faster than native apps.

Another challenge is that hybrid apps need to reproduce all the artifacts of behavior that an IoT device exhibits locally. For example, NovaSignal App (discussed in Chap. 7) removed static noise in the ultrasound data using a different software driver than was on the NovaGuide device. The result was that a sonographer conducting a patient's exam heard different sounds than the physician listening to the playback via the Cloud application, potentially leading to incorrect diagnosis. The solution was to use the same playback driver on both the device and in the Cloud application.

10.8 Summary

A hybrid application refers to the combination of on-device software with a private or public Cloud. These applications enable data and applications to be shared between a device and the Cloud. The benefits in using hybrid mode of operations include better control of data that cannot be shared outside of an enterprise boundary due to PHI or PII concerns. Additionally, it offers an ability for the users to work in an offline mode. However, the architecture of a hybrid application needs to be carefully designed to address performance and maintenance concerns. On device part of the hybrid application has the option to use a web browser or native device functionality.

Points to Ponder

1. How can one improve hybrid applications' performance and support for IoT?
2. What kinds of attacks are possible using IoT and edge devices?
3. What is the trust and security model for hybrid applications?
4. Are we at the cusp of hybrid applications with the confluence of AI and Cloud?

Answers

1. How can one improve hybrid applications' performance and support for IoT?
This can be achieved in multiple ways, some of which are listed below:

   (a) By having distributed and redundant systems for a failsafe solution, avoids a single point of failure.
   (b) Backend Cloud services are needed to log data and results for audit and machine learning inferences.
   (c) Sensors can generate enormous data requiring Cloud storage and compute power. However, moving data in and out of Cloud is slow and expensive. So, input-output considerations will require local compute and storage power.

2. What kinds of attacks are possible using IoT and edge devices?

   (a) It has been shown that an army of botnets (a term used for devices on the Internet) can be hijacked by hackers and used for launching DDOS (Distributed Denial of Service) attacks on the Cloud part of the hybrid applications.

(b) An example is of home surveillance cameras that had unsecured IP addresses used for bringing down a security journalist's blog site mentioned below: https://krebsonsecurity.com/2016/09/krebsonsecurity-hit-with-record-ddos/

3. What is the trust and security model for hybrid applications?

    (a) Security and trust models for hybrid applications need to account for local vulnerabilities for the devices.
    (b) Cloud services are needed to log data and results for audit and machine learning inferences.
    (c) Devices in the field need to trust the Cloud servers, and Cloud needs to trust the incoming device data.
    (d) Multifactor authentication offers a good defence strategy. Having any single piece of hardware or software as the sole root of trust is risky. An attacker will need to simultaneously compromise multiple devices, which is harder to accomplish than altering any single root of trust.

4. Are we at the cusp of hybrid applications with the confluence of AI and Cloud?

    (a) Yes, the architecture of hybrid applications combines the best that a device can offer in terms of partitioning roles and responsibilities: Local (for data collection and inference capabilities), and Global (for large data analysis and AI training capabilities in the Cloud).
    (b) Partitioning of data between local and global sites offers a capability for differential privacy. As a result, sensitive PHI and PII data need not be shared in a public Cloud.
    (c) A local server in an on-premises Cloud, such as in a hospital IT datacenter, allows a global AI model to be fine-tuned to the local needs based on the patient demographics.

References

1. Sehgal, N. K., et al. (2023). *Cloud computing with security and scalability*. Springer. https://link.springer.com/book/10.1007/978-3-031-07242-0
2. https://investor.hcahealthcare.com/news/news-details/2023/HCA-Healthcare-Collaborates-With-Google-Cloud-to-Bring-Generative-AI-to-Hospitals/default.aspx
3. https://www.vmware.com/topics/cloud-bursting
4. https://www.ibm.com/topics/hybrid-cloud-management
5. https://www.hhs.gov/hipaa/for-professionals/privacy/laws-regulations/index.html
6. https://gdpr-info.eu/
7. https://cloud.google.com/learn/advantages-of-cloud-computing
8. https://www.westmonroe.com/perspectives/in-brief/mobile-client-architecture-native-vs-hybrid-vs-web-app
9. Florian, R., André, N., & Andreas, S. (2014). *Towards a mobile application performance benchmark*.
10. https://www.forbes.com/sites/ciocentral/2013/01/23/html5-vs-native-mobile-apps-myths-and-misconceptions/

11. https://youteam.io/blog/high-performance-apps-you-never-knew-were-hybrid/
12. https://medium.com/@pandeyarpit88/building-a-hybrid-architecture-harnessing-the-power-of-microservices-and-monoliths-80f79b310d75
13. https://en.wikipedia.org/wiki/Conway%27s_law
14. https://wallet.google.com/wallet/u/0/home
15. https://www.apple.com/apple-pay/
16. https://medium.com/mobilepeople/a-deep-dive-into-google-pay-and-apple-pay-d56dab7194a0
17. Rivest, R. L., Shamir, A., & Adleman, L. (1978). A method for obtaining digital signatures and public-key cryptosystems. *ACM Communications, 21*, 120–126.
18. Advanced Encryption Standard (AES) (FIPS PUB 197). Federal Information Processing Standards Publication 197 November 26, 2001.
19. FIPS PUB 202. (2015, August). *SHA-3 Standard: Permutation-based hash and extendable-output functions FIPS PUB 202*. https://doi.org/10.6028/NIST.FIPS.202
20. Schneier, B. (1996). *Applied cryptography second edition: Protocols, algorithms, and source code in C*. Wiley.
21. https://medium.com/@elitemindz079/overcoming-common-challenges-in-hybrid-app-development
22. https://brainhub.eu/library/hybrid-mobile-app-development

Chapter 11
Advisory Notes for Success in Cloud and IoT Projects

11.1 Introduction

"A success" in the context of a project is defined as the accomplishment of an aim or purpose. In our setting, it signifies the completion of Cloud- or IoT-based software project on time, within its budget and to the customers' satisfaction.

This chapter begins with notes on the benefits that accrue from data-driven project management, and the design for serviceability. Next, we explore common pitfalls while executing a Cloud- or IoT-based project. This will be followed by a section on some advisory notes and a description of how to respond to challenges in migrating an existing software or IoT project to Cloud-based operations. We wrap up this chapter by briefly revisiting and reinforcing modality of the Agile process.

11.2 Data-Driven Project Management

Data-driven project management and the tools that enable this methodology are covered in Chap. 7. When implementing data-driven project management, adoption and acceptance from all the stakeholders involved allow for a higher chance of success. Since data-driven project management involves time tracking for team members, asset location monitoring, and real-time progress reports, it can be seen as intrusive by the team. As such, adoption requires alignment across the entire organization.

At the top, an organizational culture of transparency is required for data-driven project management. When all the data collected to support project, then the execution is visible to all team members as opposed to managers or executives only, everyone on the project can feel that they are supporting a collective effort to improve efficiency. Transparency also allows anyone, from the most senior

P. C. P. Bhatt et al., *Project Management in Cloud and IoT Applications*,
https://doi.org/10.1007/978-3-031-99427-2_11

managers to the entry-level engineers, to identify areas for improvement. Any self-reported data also benefits from transparency since those data will be visible to peers and subject to peer-review.

Among the team, a champion for data-driven project management is beneficial. The champion can be any individual contributor(s) in the team. For example, the champion may be a manager, a software developer, a verification engineer, etc. The key attribute of the champion is that they constantly express support and promote use of the tools for data-driven project management. A champion may not need to be told to express this support but will inherently understand the value of the method and want to share the knowledge and insights. When a team member enthusiastically participates and supports a cause, others will notice. This can be the basis for building support across an entire team.

Agile principles have been applied across many domains, not just software development [1]. In the context of data-driven project management, iterative deployment of tools and processes can lead to an improved outcome. Time tracking software can have many features. Project and resource planning tools are forms of hybrid database-spreadsheets and are designed for configurability. It can be overwhelming to try and use all features or design and deploy the perfect resource management system to an internal team all at once. Instead, it is more reasonable to deploy the data-driven project management tools with minimum functionality at first. Once a project team has had a chance to interact with the tools and provide feedback, further deployments can be scheduled to increase functionality and adoption.

In addition, serviceability is a worthwhile consideration to ensure long-term success of software products.

11.3 Serviceability of Software Products

Cloud- and IoT-based software projects offer a unique opportunity to amplify the effectiveness of Design for Serviceability (DFS). With Cloud connectivity, hardware and software projects can upload errors, usage, and service logs to be stored in the Cloud. If a software product is well designed to leave these breadcrumbs, or clues, the log holds a treasure trove of very valuable information. Product designers can use this information to develop and release software patches to fix products in the field. The information can also be used as input to future product features. For products that are not Cloud connected, it is important to request the customers to share their logs or, in the worst case, service technicians to visit the customer sites and manually download these logs. A significant operational advantage exists for Cloud-connected projects by utilizing the "always-connected" nature of the Cloud software (or application).

With logs uploaded to Cloud, its provider company has all the data needed to analyze complex issues. Developers can build tools and dashboards to share product usage information. This information can be shared both "up" and "across." Sharing "up" allows senior management to easily understand issues in the field. Sharing "across" allows developers to see how their product is being used in the field and what are the new use case scenarios they did not think of. Whether sharing "up" or "across," the result of analyzing logs uploaded to the Cloud provides a path to continuous product improvements.

Once logs have been analyzed and issues resolved, the company is poised to offer updates to customers. An advantage of fully Cloud-based software is that patches can be made in the Cloud and applied with no effort on the customer's part. For hardware projects that have a Cloud-connected software component, that software may be able to push patches to the hardware through the Cloud interface(s). This process saves the company time and money by avoiding a lengthy and expensive on-site patch rollout.

Incorporating DFS, particularly in the early stages of a product's life cycle, leads to increased value to the customers and service departments of the companies developing Cloud-based SW. By resolving issues quickly, customer satisfaction remains high. This can result in favorable public online reviews and word of mouth messages in the industry. The net outcome may help increase in profitability of the company.

11.4 Pitfalls to Avoid

Many software projects fail to deliver on schedule or within the allocated budget due to poor estimation of work, frequent changes of requirements or sometimes not having the right people for the tasks. According to Standish Group's Chaos report, more than 60% of software projects fail [2].

Below are some common pitfalls that may impact a Cloud- or IoT-based software project:

1. **An unrealistic development schedule:** If enough time is not allocated to develop the core features, then a project is likely to fail. In addition to that, if and when, the System Requirement Specifications (SRS) or design strategy changes, the Cloud- or IoT-based project starts to fall behind. This may result in poor implementation decisions made in a rush.
2. **Resource Scarcity:** Note that both the quantity and quality of resources matter. Not having sufficiently experienced developers would result in re-inventing the wheel. This not only increases the amount of work to be done with limited resources but also results in imbalance of work and results in schedule slips.
3. **Underestimating a project's complexity:** Generally, functional requirements may be well understood, but a lack of clarity on the cross-platform support, performance and scalability requirements increase the technical challenges for the development team. If these are not well anticipated in advance, then technical issues will get out of control causing project failures.
4. **Overdependence on third-party vendors:** By using third-party software components, one may get ahead and reduce complexity. But if proprietary tech-

nologies are used and one of the external components slips in delivery schedule, then that would have cascading consequences for the main project.
5. **Budget Constraints:** A Cloud vendor may change their pricing models causing the cost assumptions of main project to go completely haywire.

In addition, IoT projects are susceptible to sensors, devices, or control chip obsolescence [3]. Sometimes, IoT devices are located in places that cannot be easily reached, for example, multi-story high telecom towers or windmills in a hilly terrain. Also, devices may get recalled for security fixes but in some cases that would be hard to do as in the case of pacemakers installed in patients' chests. Another example is of smart cars with IoT systems that may reach the end of their useful life long before the vehicle's end of life. A potential solution in this case is a modular design, so users can replace the functional parts, like changing the battery in a cell phone. Another solution is to use software patches to handle security issues with IoT devices using driver upgrades. As a foresight this may be achieved by providing some extra non-volatile on-chip memory in the IoT device.

Similarly, a Cloud datacenter may contain servers, routers, and other equipment from different vendors and product generations under the same roof. This enables IT managers to upgrade their hardware in phases. Then orchestration software takes care of scheduling appropriate tasks on the matching hardware, such as AI/ML software on GPU-enabled servers.

11.5 Cloud-Based Software Considerations

One often cites the Manhattan project [4] as an iconic benchmark for large-sized projects. In the absence of computers, engineering of Manhattan was accomplished using pencil and reams of paper [5]. Something akin to this started happening in software designs. Software projects became bigger in size and complexity. North Atlantic Treaty Organization (NATO) in 1950–70 was the single largest promoter of software contracts. NATO recognized that software development, unlike other engineering projects, lacked in known or set processes and standards. NATO needed guarantees from vendors. It was imperative to be accountable to enhance the success rate and minimize the failure of software projects. NATO, therefore, promoted efforts in evolving definitive methods to direct software development work. This helped in identifying various phases of software development with much better estimates of time and effort for each phase. To stay aligned to the general engineering practices, NATO coined a term "Software Engineering" [5] to broadly cover the software projects that involved human activities spread over many man-months. A classical book by Fred Brooks entitled "The Mythical Man Month" [6] best captures the dilemmas of Software Engineering.

A key component of a Cloud-based project is its software design. In this section, we briefly deal with the basic concerns and methodologies that have evolved to help in the design of large systems.

The modeling frameworks discussed in earlier chapters can help software system designers to evolve a robust, reliable, and maintainable design strategy. Software designers and architects identify functional modules much in the same way as civil architects designing homes do. Just as the civil architects identify functional areas like kitchen, bathrooms, living room, study, etc., software architects segregate key functional segments and work out the details for each segment to finally compose the whole artifact.

In Cloud software design context, modularization was best explained by David Parnas [7] by enunciating the principle of separation of concerns. The principle of separation of concerns helps a designer to identify modules to finally compose a design. The methodology requires that the overall goal is divided into sub-goals—each realized by a separate module. The synergy in operation requires communication among these modules. The decomposition into modules is done very carefully to ensure that each module is as independent as may be feasible in meeting the sub-goal(s). The sub-goals are dictated by the design requirements. Modules communicate primarily to achieve the system design goals. Such modules are considered to have high cohesion and minimal coupling [8]. Cloud software system modularization, based on the principle of separation of concerns, aligns naturally and effectively with OO (Object-Oriented) thinking. It relates to using hierarchical relationships between objects, each of which may encapsulate some data elements and procedures (aka methods) to manipulate them. External entities can access these objects only via pre-defined methods. Let us consider an example from the transportation sector to define a class of objects, such as cars. Every instance of the class cars is an object with four wheels, an engine, some form of steering and power transmission mechanism. Within this object, a wheel may be its data element with details such as the tire type, tread depth and air pressure, etc. But the complexity of such details is hidden from the user of a car, unless they call a method to check its tire pressure. Note that a sub-class called an electrical car can be derived (aka inheritance), with an electrical motor engine using a battery and details of its charging. Similarly, a second sub-class can be a gasoline car with its own details, e.g., number of cylinders. Such alternate behaviors may be captured by invoking polymorphism, e.g., referring to multiple objects with a single name. Thus, encapsulation, polymorphism, and inheritance collectively form OO thinking. It ensures greater maintainability, reusability, and extensibility [8].

Every Cloud project needs to consider the following aspects:

1. Reliability.
2. Maintainability.
3. Scalability.
4. Modifiability.
5. Standard APIs.
6. Performance.

A practical aspect of Cloud-based software design relates to the ease of interaction with human users. This is variously regarded as Human Computer Interaction (HCI or CHI). Whenever we use an application (App), we interact with the App using a User Interface (UI) offered by the software designer. Besides the UI, applications need to communicate with other applications. In general, therefore, there may be internal communication:

(i) Between program modules within an application.
(ii) At higher level between applications. This form of communication has resulted in a few new design considerations—the design of an application program interface (API).
(iii) In other words, UIs and APIs need to be crafted carefully. APIs ought to be embedded in the design upfront and should not be attempted to retrofit.

11.6 Additional Challenges in Migrating to Cloud

In addition, migrating a software- or IoT-based project to Cloud brings additional challenges [9] as outlined below:

1. **Lack of executive level buy-in:** The mindset of individuals at executive level matters both in the product development company and on the customers' side. Some people think that Cloud is free, open-source, and easy to implement, which is not true.
2. **Too much executive level buy-in:** Some leaders are overly optimistic and enthusiastic about Cloud and may take too great a leap in a short time. This approach is also known as "lift-and-shift," means one can take old existing processes and products, and simply migrate them to Cloud. Workflows need to be re-evaluated and recreated using new Cloud-based tools and services.
3. **Lack of Leadership:** Sometimes it is not clear who is responsible for making decisions about which Cloud and which services to use. The expectations and accountability need to match with clear communications and tracking milestones set in place. The leadership ought to work with that kind of telescopic foresight.
4. **Integration skills:** A new Cloud- or IoT-based product needs to often interface with existing in-house software products. This needs to be clearly defined and tested. Else it will become an iterative learning process causing cost overruns and schedule delays.
5. **Secure by Design:** Products designed with "Secure by Design" principles prioritize the security of customers as a core business requirement, rather than merely treating it as a technical feature [10]. During the design phase of a product's development life cycle, companies should implement Secure by Design principles to significantly decrease the number of exploitable flaws before introducing them to the market for widespread use or consumption. Out-of-the-box, products should be secured with additional security features such as multi-factor authentication (MFA), logging, and single sign-on (SSO) available at no extra cost. Products need to be tested for API security with end-to-end use cases. IoT devices can benefit from Pentesting with a specific focus on firmware and com-

munication hacking and a radio security assessment [11, 12]. It should be noted that the weakest link in the chain may break, compromising confidentiality, data integrity, or availability during the product deployment and on-field usage.

6. **Design for Performance:** This is the practice of creating systems that are efficient, responsive, and fast. The goal is for an IoT and Cloud application to handle increasing workloads without perceivable performance degradation. IoT analytics predicts that in 2027 there will be more than 29 billion IoT connections [13]. According to Statista, the IoT sensors market is expected to reach 43 billion USD in 2025. Network latency between IoT devices and the Cloud can affect real-time applications. Using a globally distributed Cloud platform can help to reduce this problem. Key strategies for scalability [14] and performance include:
   (a) Microservices architecture: Scale individual application components.
   (b) Auto-scaling: Cloud platforms automatically adjust resources based on demand.
   (c) Serverless computing: Pay only for resources used, enabling easy scaling.
   (d) Containerization: Package applications for efficient scaling across environments.
   (e) Offloading: Transfer resource intensive computations from constrained embedded platforms to resource-rich platforms such as the Cloud [15].
   (f) Adaptive power management: Carefully consider which sensors and sub-circuits are powered based on the real-time performance requirements of the device.
   (g) Link budget planning: Optimize the wireless protocol and circuitry specifications to ensure sufficient RF signal strength for the application [16].

11.7 Revisiting Agile Processes

A few facets of Agile [8] need to be revisited as a key to success for every IoT or Cloud-based software project. These are:

1. **Sprint:** Agile process requires that the project work is divided into multiple stages to achieve a prescribed time-based sprint. Each stage is expected to achieve a small but functionally operating release. Each stage of the process requires that ownership is undertaken to meet the sub-goals set for that stage. Furthermore, releases are tested fully to qualify to be taken forward to the next stage. Time boxed sprints help in maintaining the momentum of progress.
2. **Manageable units:** The idea of "Sprint" is primarily to be able to split the project parts to small manageable units which can be closely monitored and tested. This helps immensely in mid-course corrections. Errors are caught and fixed in the stage where they occur. This mitigates the curse of cost and effort amplification observed in Waterfall model. Smaller manageable units help in having predictable time and effort estimates as well.

3. **Ownership:** Every stakeholder is involved in every stage of the development. This means that end-use client(s), business promoter(s), business development manager, software architects/ designers, coders, software testers, and maintenance folks all participate in taking the process forward. This is achieved through Scrum [8] reviews. Of course, the ownership is taken by each relevant group to move the project forward to meet the stated requirements.
4. **Testing:** Testing is not an afterthought or the very last stage in Agile. Agility requires embedding testing as a part of the process at each stage. Ideally tests are pre-designed, and the designers and coders need to qualify over test thresholds before entering the next stage of the process.

Let us quickly see the connection between Agility and separation of concerns design principle and also how Agility connects with object-oriented thinking.

11.8 Separation of Concerns

The entire exercise of Cloud- or IoT-based software design ought to be thought of in macro and micro terms. Macro obviously refers to thinking about a large system composed of modules, or IoT subsystems, without being distracted by their internal details. This is precisely the top-down design methodology. On the other hand, micro considerations entail a bottom-up approach. Within a module, the identification of key objects and classes is warranted to ensure that the use-cases are covered. Designers need to ensure this in the detailed design with code and interface options.

For macro level design, Agility brings in the three important facets:

1. How the modules or subsystems may be sized
2. How the modules or subsystems may interact
3. How the workload may be split with the resources available

Micro level design presents a thought process that is locally processed. The basic idea is to provide specific services to other modules. Agile thinking complements the separation of concerns design methodology as well as OO thinking to that extent.

The primary objective is to achieve the goals initially stated in the requirements. With Agile processes requiring embedding the testing at every step, the course correction is relatively easier than for the Waterfall model. Also, heavy duty Cloud-based software testing activity happens during micro level development. More details on the software design methodology and processes are covered in [8].

11.9 Summary

In this chapter, we revisited data-driven project management besides introducing design for Serviceability (DFS). We also explored common pitfalls to avoid while executing a Cloud- or IoT-based software project. This was followed by a description of challenges in migrating an existing software or IoT project to Cloud-based operations, and proposed responses. We wrapped up this chapter by briefly revisiting and reinforcing modality of the Agile process.

Points to Ponder

1. Does Agile methodology of program development support, complement the following concepts:
   (a) Stepwise refinement
   (b) Separation of concerns
   (c) OO thinking
2. Would one be right in thinking of flowcharts as high-level visual program structures?
3. Are there any security concerns with sharing and storing customer logs under DFS?
4. What are some benefits to Design for Service?
5. Why is transparency important for data-driven methodology to succeed?

Answers

1. Does Agile methodology of program development support or complement the concepts of stepwise refinement, separation of concerns, and OO thinking?
   (a) Agile methodology can be deployed in addition to stepwise -refinement, separation of concerns, and OO thinking. It is an approach that is most useful when not all the user requirements are known or fixed at the beginning.
   (b) A stepwise refinement is an iterative approach, which can be used to fine-tune a solution over time. An example can be to speed up an algorithm or add more details in the UI (User Interface) of a program.
   (c) Separation of concerns is a principle used in programming to separate an application into units, with minimal overlapping between the functions of the individual units. It can be obtained using modular or layered approach, or by identifying the hierarchy in the system.
   (d) Lastly, OO (Object-Oriented) thinking relates to a hierarchical relationship between classes and objects, each of which encapsulates some data elements and procedures (aka methods) to manipulate them. External entities can access these objects only by using very specific methods.
   (e) In using Agile methodology, one can use one or more of the above techniques. It is essentially a way to manage a complex project by breaking it into several phases. The work cycles through a process of planning, execution, continuous testing, and evaluation. Agile requires constant collaboration between stakeholders and the outcome results in continuous improvements at every stage.
2. Would one be right in thinking of flowcharts as high-level visual program structures?

(a) Yes. A flow chart represents a series of actions performed in the order of time as the events occur. The decisions are made to direct the program control flows through the steps in data processing.

3. Are there any security concerns with sharing and storing customer logs under DFS?

   (a) Yes. If customer logs contain PHI (Private Health Information) or PII (Personal Identifiable Information), then it needs to be removed or sanitized before storing them. If for any reason PHI or PII information needs to be stored, then it should be HIPPA compliant in the USA, GDPR compliant in the EU, and corresponding local laws in other countries.

4. What are some benefits to Design for Service?

   (a) Design for Service can increase value to both customers and design teams. Customer satisfaction can be increased with quick issue resolution, and product teams are exposed to user-centric issues with the design.
   (b) When customer satisfaction is high, product reviews are likely to be favorable. Outstanding reviews and word of mouth can lead to increased sales for the product and profitability for the company.

5. Why is transparency important for data-driven methodology to succeed?

   (a) An organizational culture of transparency is not just important, but essential for data-driven project management.
   (b) When data collected to support a project is visible to all team members, everyone on the project participates in a collective effort to improve efficiency.
   (c) Transparency also allows anyone on the team to identify areas for improvement.
   (d) Any self-reported data also benefits from transparency since those data will be visible to peers and subject to peer-review.

References

1. Pikkarainen, M., Salo, O., & Still, J. (2005). Deploying Agile practices in organizations: A case study. *EuroSPI 2005, LNCS, 3792*, 16–27. https://doi.org/10.1007/11586012_3
2. https://www.standishgroup.com/news/45
3. https://www.networkworld.com/article/965848/iot-has-an-obsolescence-problem.html
4. https://www.history.com/topics/the-manhattan-project
5. https://ifs.host.cs.st-andrews.ac.uk/Books/SE9/Web/History/index.html
6. https://en.wikipedia.org/wiki/The_Mythical_Man-Month
7. https://effectivesoftwaredesign.com/2012/02/05/separation-of-concerns/
8. Bhatt, P. C. P. (2021). *Software design, architecture and engineering: Concepts and practice*. Prentice Hall India, Easter Economy Edition. ISBN-978-81-951611-4-0.
9. https://cloudacademy.com/blog/the-top-5-pitfalls-of-cloud-migration-and-how-to-avoid-them/
10. https://www.cisa.gov/securebydesign

11. https://www.iotpentestingguide.com/
12. Guzman, A., & Gupta, A. (2017). *IoT Penetration Testing Cookbook: Identify vulnerabilities and secure your smart devices*. Packt Publishing.
13. https://www.techaheadcorp.com/blog/a-guide-to-cloud-infrastructures-in-the-iot-app-development/
14. Sehgal, N. K., et al. (2023). *Cloud computing with security and scalability*. Springer. https://link.springer.com/book/10.1007/978-3-031-07242-0
15. Čolaković, A. (2023). IoT systems modeling and performance evaluation. *Computer Science Review, 50*, 100598. ISSN 1574-0137. https://doi.org/10.1016/j.cosrev.2023.100598
16. https://resources.altium.com/p/rf-signal-chain-and-link-budget-basics-iot

Chapter 12
New Value Creation

12.1 Cloud-Based Opportunities

The advent of Cloud computing [1] has opened many opportunities for new value creation. These largely fall into two categories. One is to create new products that reduce a customer's Total Cost of Ownership (TCO) and offer new services. The second category relates to in-house migration of existing software products and services to Cloud.

In the latter category, business models also evolve from product ownership to periodic subscription payments by the users. In both of these categories, new winners emerge in the marketplace as customers migrate to adopt Cloud-based applications.

This chapter describes three examples, one based on PaaS and the other two based on SaaS, to demonstrate how one may utilize Cloud-based operations to offer enhanced value propositions. We shall explore the basic architecture of these systems and also demonstrate how collaborating team members spread across geographies benefit from the capabilities offered by product suites.

The first example we explore is that of Salesforce [2], which created a Cloud-based CRM (Customer Relationship Management) tool, with a web browser as its user interface. It flew in the face of a problem that the majority of business people are distrustful of a third party holding their valuable customer contract data. Salesforce, as a company, needed to convince buyers that their customer data was secure from loss and/or intrusion.

The second example we shall explore is Microsoft's Office suite migration [3] to Cloud. In the past decade, Microsoft's revenue growth stalled as personal computer refresh cycles were getting annoyingly prolonged, impacting revenue. The company tried to challenge Amazon in Cloud but was not successful as Amazon's AWS continued to dominate the public Cloud marketplace in the USA. Recognizing the ground reality, Microsoft management came up with the idea to migrate its popular

P. C. P. Bhatt et al., *Project Management in Cloud and IoT Applications*,
https://doi.org/10.1007/978-3-031-99427-2_12

office suite to MS Azure Cloud. That was a master stroke as it readily offered Microsoft an existing installed base of users, in the retail and enterprise domains. This move also encouraged enterprises to put their other business data in the Azure Cloud, making MS the second largest Cloud player in North America.

The third example that we describe here offers an insight into Netflix's switch of business model for distribution of their content [4]. Netflix started its operations in 1998, by mailing out physical CDs and DVDs of movies, shows, and video games through US Postal service. As Internet technologies evolved, Netflix tweaked its business model enabling customers to stream their favorite contents anytime, anywhere. The switch in the business model catapulted it to become world's ninth-largest Internet company [5] by revenue in 2023. It should be remarked that Netflix's success can be attributed to its evolution into Cloud-based operation, which facilitated distribution of streaming media products across multiple geographies.

These examples are in addition to the NovaSignal startup that we studied in an earlier chapter. This medical device company has successfully added a software application to its ultrasound device for storing and accessing brain blood flow data in the remote Cloud portals.

We conclude this chapter by describing the use of AI, including an emerging product called ChatGPT and evolution of domain-specific Small Language Models (SLMs). Digital Twins (DT) and their relations to IoT sensors and products are also described.

12.2 A Humble Beginning

Salesforce began in a small apartment in San Francisco during 1999, just before the dot-com boom cycle ended in a bust. Its co-founder Marc Benioff (encouraged by funding by Larry Ellison, Oracle's CEO at that time), left a lucrative sales career at Oracle to start a new Cloud-based SW offering. Within 10 years of its founding, Salesforce was able to reach $1 billion in revenue with over 55,000 customers in the CRM domain where Oracle was once the leader [2].

Currently, Salesforce is one of the largest technology companies in the world. As of September 2022, it had a market capitalization value of $153 billion and it is ranked as 61st largest company in the world by revenue [6]. Its products include several CRM technologies, such as Sales Cloud, Service Cloud, Marketing Cloud, and Commerce Cloud and Platform. Additional product offerings are tools that enable customers to create apps, integrate data from other systems, visualize data, and offer training courses.

Salesforce offerings use Platform as a Service (PaaS) that allows developers to create add-on applications, which integrate into the main salesforce.com application. These third-party applications are hosted on salesforce.com's infrastructure in the Cloud. These applications are built using declarative tools, using proprietary Java-like language, as well as XML to finally generate HTML. There is an online application marketplace, called Salesforce AppExchange, that connects customers with third-party applications and consulting services. As of 2021, this exchange had over 5000 applications listed and counting.

12.3 Salesforce Architecture Fundamentals

Salesforce Customer 360 Platform [7] is based on a multitenant, metadata-driven architecture to every individual Salesforce tenant instance (aka "org"). There are three key aspects of this platform:

1. **Transactions:** A transaction can be instantiated by code execution or a database operation. The platform sets governance and execution limits to ensure that no individual transaction can monopolize any of the shared compute resources. A built-in order of execution prescribes how an organization's configuration and code will behave. Results are not committed, nor data transactions are finalized, until the final "after trigger" behavior as described in the order of execution has completed successfully. This conforms to principle of serializability in distributed systems. The operation should cover all the workflow and escalation rules. After their execution, system validation steps are executed. This helps to verify that all the required fields have a non-null value, to ensure consistency. A fatal error or disqualifying behavior during a data transaction will cause the platform to roll back all the data operations. This ensures data integrity (as depicted in Fig. 12.1) for every transaction's execution flow.

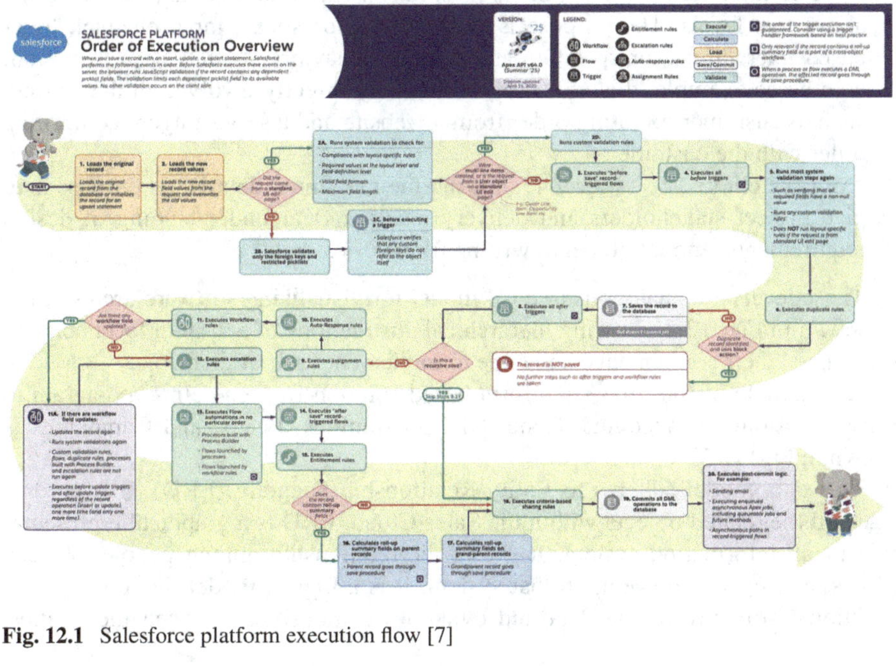

Fig. 12.1 Salesforce platform execution flow [7]

2. **Data and Metadata:** It is important to understand what is considered as data and metadata from the platform point of view. Below are some examples:
   (a) Data copied from production into sandbox environment: metadata.
   (b) Migrated by direct upload load into an application: data.
   (c) Migrated by Metadata APIs: metadata.
   (d) Individual records in a Salesforce org: data.
3. **Platform APIs:** There are many Salesforce platform APIs to support different data formats and protocols, various operation types, and timings. These are built to handle large transaction volumes. Salesforce platform supports backward compatibility, which means a particular version shall perform with stability and consistency, even as new API versions are released.

The above platform enables a customer's marketing, strategy, or sales delivery teams to execute a campaign. This is accomplished by "Diagrams," which is a set of steps to walk through from an idea to the execution details. Below are the 4 levels for Salesforce diagrams:

The big picture: This is level 1, showing the "art of the possible," or a digital transformation scenario. The purpose is to show business capability and high-level product positioning, without getting into many details. The focus of level 1 is to clearly convey the big picture with no overwhelming details.

Piece of the whole: This is level 2, showing a single business capability or solution overview. The purpose is to show specific technology or product subset, and some more details about system capabilities or relationships between entities.

Process or interaction view: This is level 3, showing finer details at either of the above two levels. The purpose is to show user personas, input channel, time-based interactions, or specific details about behaviors at each step. In a retail commerce example, this will show technology directly involved in the process such as customer initiating order from a website and a service agent confirming order with the customer.

The double click: This is level 4, showing the maximum details. Its purpose is to give project stakeholders and delivery teams a clear understanding of design requirements, enough to begin writing user stories.

These levels are analogous to a C4 model for visualizing software architecture, as shown in Fig. 12.2, showing hierarchical abstractions of a system level context, containers, their components, and code to implement each of them.

An example of Salesforce platform-based solution for generating a diagram to track a customer's Accounts, Contacts, Opportunities, Leads, and Campaigns is shown in Fig. 12.3.

This data model adheres to Entity Relationship Diagram (ERD). It is used to track all the related objects within the Sales Cloud. ERD is a graphical representation of an information system. It shows the relationship among people, objects, places, concepts, and events in that system. It is a logical model that conveys the functional structure of the data and dynamic relationships between the entities.

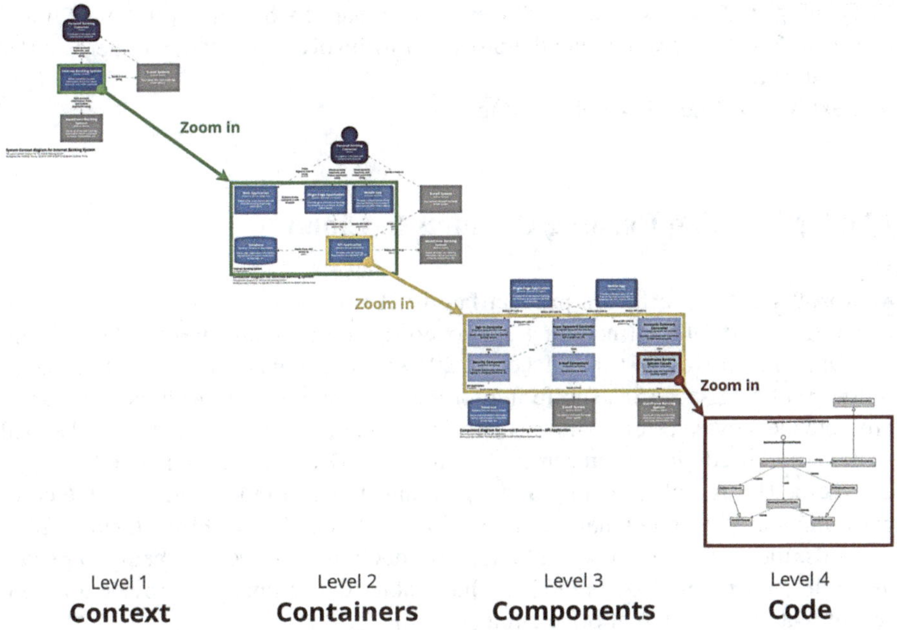

Fig. 12.2 Hierarchical abstractions using C4 model [7]

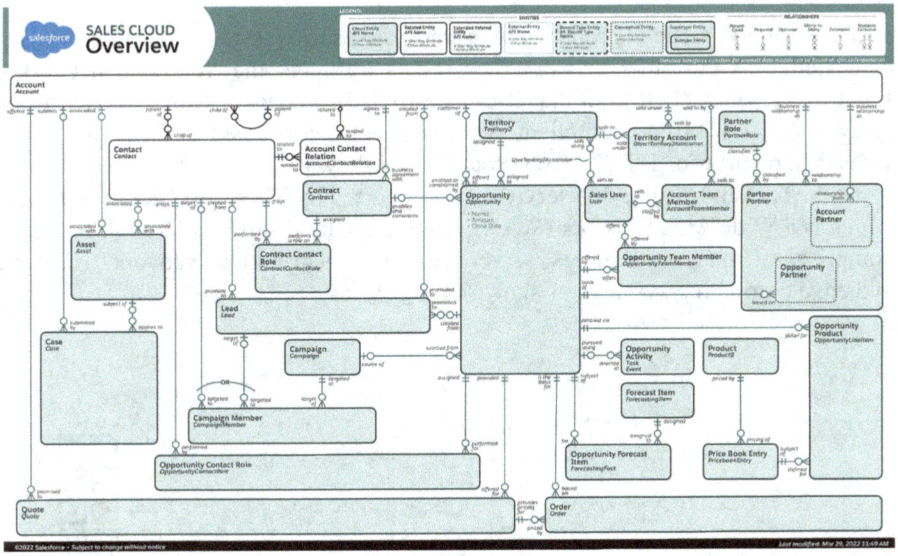

Fig. 12.3 Salesforce platform execution flow [7]

ERD in Fig. 12.3 shows a structural view, which maps to the temporal flow of information in Fig. 12.1, by showing the process flow involved in retail commerce order management.

Next, we explore our second example.

12.4 Migrating Existing Products to Cloud

Microsoft's 365 is, in fact, a product family. It was formerly marketed under the name Microsoft Office (including applications such as Word, Excel, PowerPoint, and Outlook). It was first introduced in 2010 as a reference to subscription-based Software as a Service (SaaS) offering for the corporate market. Microsoft began to promote the service as the primary distribution model for its office suite and added customer-focused plans with services such as OneDrive (for Cloud-based storage) and Skype (for remote meetings and communications). In October 2022, the company announced discontinuation of "Microsoft Office" brand. From January 2023 onwards, most of its products and online productivity services are being marketed under the "Microsoft 365" brand [3]. This enables the company to provide different sets of features at different price points.

The Office 365 SaaS solution can work for individual customers as well as large enterprises. It combines dozens of different products, as shown on the right side of Fig. 12.4. Office 365 supports multiple TCP sessions for every application for each user.

The network traffic is protected by a stack of security systems at the customer's egress edge. A customer network then routes traffic to the Microsoft network's Point of Presence (POP), so that it can be routed to Azure's Cloud service front gateway [8]. For Office 365 client systems directly connected to the Azure's services, Microsoft uses HTTPS web protocols, as shown in Fig. 12.5.

At the backend, Office 365 services are supported by Microsoft Global Network, which is a Wide Area Network (WAN) providing Cloud services through Azure. It comprises of many datacenters, servers, Content Distribution Networks (CDNs), and edge computing systems, as shown in Fig. 12.6.

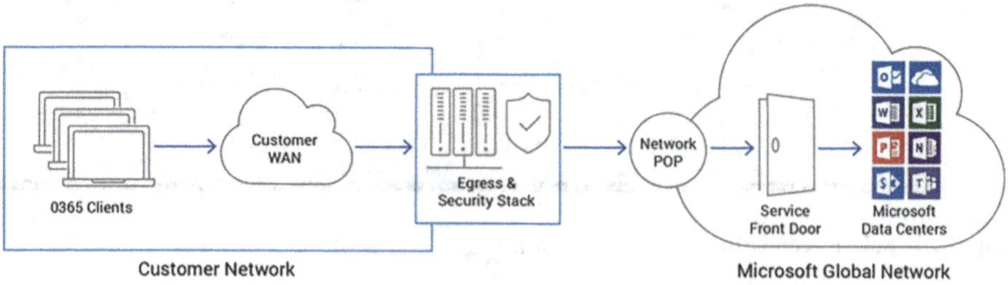

Fig. 12.4 Microsoft's Office 365 solution architecture [8]

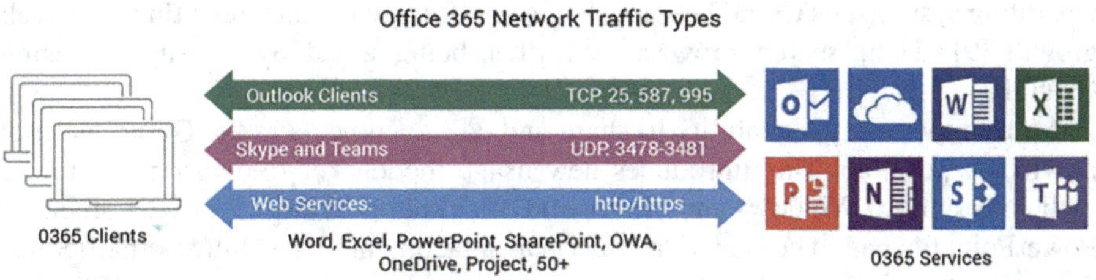

Fig. 12.5 Microsoft Office 365 clients connecting to Azure Cloud-based services [8]

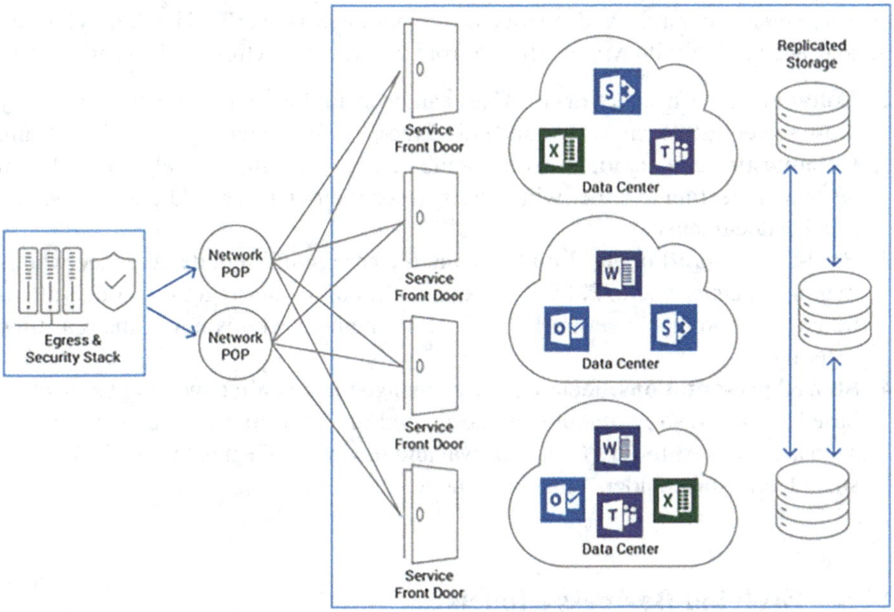

Fig. 12.6 Microsoft Global Network to support Office 365 [8]

In summary, Microsoft has expanded its Cloud infrastructure to scale and support office applications for millions of individual users and scores of enterprise customers. This has opened up new revenue streams for the company and offers new usage models for its customers.

12.5 New Usage Models

Cloud-based product migration can help to create new companies such as Salesforce and help existing companies to generate billions of dollars in additional revenues. As Microsoft's revenue growth slows from traditional products, such as Windows

Operating systems, its Cloud business helps to offset the decline, resulting in overall growth [9]. Their search browser is further being aided by AI support using ChatGPT.

More importantly, the ability to share and edit documents in the Cloud was not previously possible. This introduces new usage models for the customers. In the earlier versions of Windows and Microsoft Office suite, if a large Word document or PowerPoint file had to be shared, it was via an attachment to the Outlook mail. Since the mailer had limitations on the size of a file, and larger file transfers took longer time, this was an inefficient method of sharing and collaboration.

Now with Cloud-based storage, only a pointer to the large file needs to be exchanged. Edits and updates by one user are easily visible to all the other users as long as version controls and history are maintained properly. The following are some usage models with Microsoft 365 from anywhere, anytime collaboration [10]:

1. **Collaborate with select peers:** Files can be shared with trusted peers, and only those select people can access and edit. Later, the files can be shared with a team.
2. **Collaborate as a team:** A team member, even though remotely located, can work in a meeting together with others to see their changes, add comments, and finalize documents.
3. **Real-time co-authoring:** Edit the same file across different devices, see who is making changes where. Put @username in the comments to get someone's attention as the comments get pushed to a team member who is not in the real-time meeting.
4. **Shared presentations:** Ideas can be exchanged in an online meeting using chat-based workflow. New people can be invited or added to a shared presentation, such as reviewers to give final approval and feedback. Then resulting files can be shared with stakeholders outside of the team.

12.6 Evolving Business Models

Netflix is one of the most successful entertainment media companies of all times [4]. It was started in 1998, to avoid late fees for return of VCR tapes and DVDs at a then brick and mortar store chain in the USA called Blockbuster. Founders of Netflix avoided the expense of physical stores and passed their savings to customers by offering physical copies of movies and shows by US mail for a fixed monthly price. As Internet technologies evolved and streaming bandwidths became available in a cost-effective manner, Netflix offered millions of genres to its online subscribers through its partners in the media businesses. Netflix value proposition is 24 × 7 streaming without interrupting advertisement for the shows in high-definition content. As of this writing, the company is responsible for 15% of Global Internet traffic [4]. This is substantially higher when compared to 3.7% for Amazon's Prime videos. This is significant considering that Netflix uses Amazon's Cloud Platform.

To optimize streaming its videos online, Netflix encodes a movie or show in different video formats, each customized to a different client device, such as 4K or 1080P TVs, a browser screen on a PC, or the small screen of a cell phone. This enables Netflix to transfer the least number of bytes to deliver consistent user experience and make its videos compatible across multiple devices. This is shown in Fig. 12.7. Most of the hard work of encoding or transcoding is done offline and upfront using Amazon's public Cloud. Subsequently, multiple copies of a movie are stored. Then an appropriate copy is picked to serve the content based on a user's device and network speed.

Netflix's architecture [11] has three parts:

1. **Client App:** Netflix has an interactive user interface (UI) that works on almost all devices, such as cell phones, tablets, PCs, and TVs. It enables users to search videos and provide recommendations based on the previous user preferences. React.js (java script) is used to write the front-end due to its loading speed and performance.
2. **Backend:** Netflix has chosen a microservices architecture [12], which enables it to structure the application as a loosely coupled collection of services that can be independently deployed in the Cloud. Netflix currently uses Amazon's AWS to serve its customers. Microservices, as shown in Fig. 12.8, allow Netflix to scale and run the following services:

   (a) User login and authentication service.
   (b) Subscription management.
   (c) Video service.
   (d) Transcode service.
   (e) Global search.

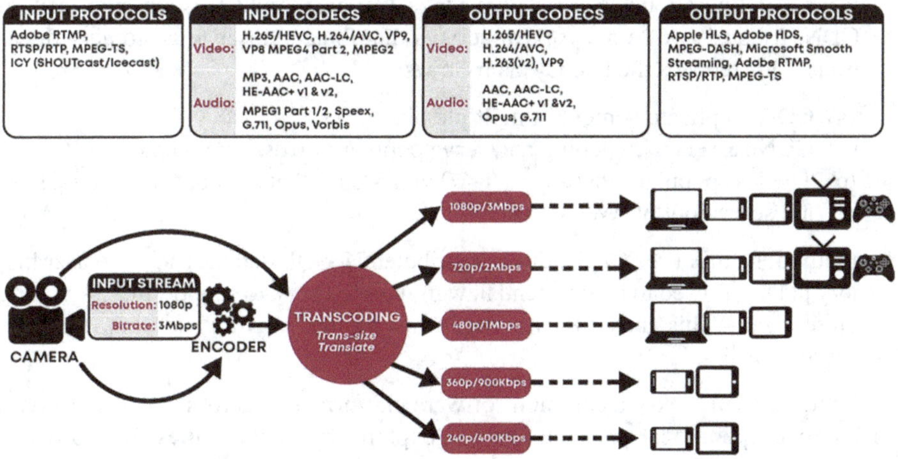

Fig. 12.7 Netflix onboarding new content [11]

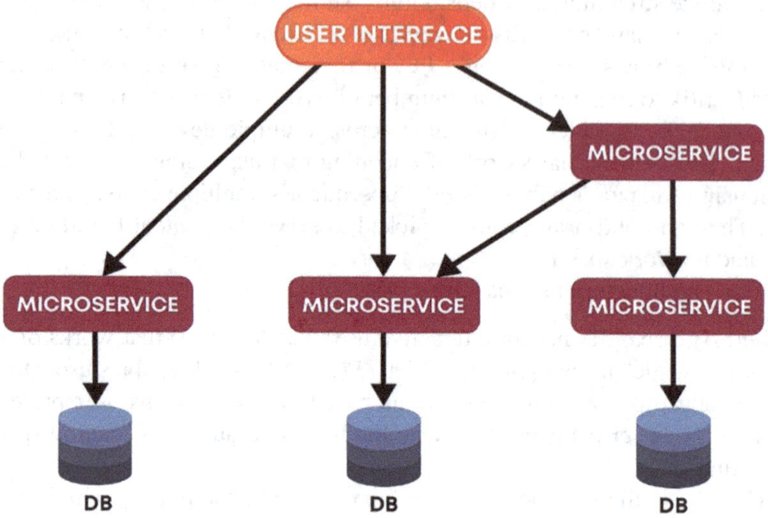

Fig. 12.8 An example of microservices [11]

There are two types of servers used by Netflix:

(i) **Critical Servers:** These run the services used frequently by users such as streaming of popular videos. These servers are kept independent of other services, so that the users can continue to watch with low latencies.
(ii) **Stateless Servers:** These serve API requests to clients and are deployed in a way so they can continue work with other instances even if a server fails. This ensures high availability.

3. **Content Delivery Network (CDN):** In parallel with storing and serving videos from a Public Cloud, Netflix uses Open Connects, which is its own custom CDN. It consists of a network of servers distributed closer to customers' locations. This reduces the latency, as follows:

   (a) CDNs replicate content in multiple places.
   (b) CDN servers use caching and serve popular videos from memory.
   (c) The less popular videos (i.e., 1–20 views/day) that are not cached by CDNs are served out of AWS datacenters.

In addition, Netflix uses Cassandra, a distributed NoSQL database to store user history [11]. It is used to recommend new movies to users, to support their watching preferences. This aids in user stickiness besides driving Netflix's business success.

Next, we shall look at a common software platform used across many enterprises to improve operational productivity and explore the opportunities to improve it using Cloud-based solutions.

12.7 A Brief Note on Improving Enterprise Resource Planning (ERP) Systems

An ERP is a broadly deployed solution in enterprises to automate the processes, improve business operational performance, and reduce cost [13]. While the basic design of most ERP systems remains product centric, new functions are being added to take advantage of underlying technological improvements. Current Cloud-based solutions are offered as SaaS by companies such as SAP, Oracle, and Salesforce.

At this juncture, we feel that a revolution is due in supply chain management. An example is the war in Ukraine impacting supply of wheat all around the world. It took several months to adapt. Similarly, airlines had to reroute their flights from the conflict zones. Note that currently all US carriers avoid Russia as a no-fly zone.

Al can help to identify potential problems in supply chain logistics before these can impact a product launch schedule. An example is that of a business that sources components from other countries. Sometimes the supplier may be experiencing labor problems or transportation delays due to global conflicts. An AI-based system can anticipate the impact and propose alternate solutions for replacement or alternate sourcing of components. This will help to meet business commitments in a timely manner. Another example is of labor shortages in one part of the country and opportunities for outsourcing, which a web-connected AI solution can explore the needed arbitrage.

Another sub-area within the ERP is to ride the trend of decentralized datacenters [1]. With geographically spread teams, data need to be mirrored across the sites to minimize latency. This also provides opportunities to maintain high availability with redundancy of servers and storage systems. It requires applications to be built using distributed Cloud networks. A distributed ERP solution will require multiple mirror sites and frequent data synchronizations to be effective.

12.8 Emerging Technologies

An important emerging area is Artificial Intelligence (AI) operations within an Agentic framework. AI in automation defines an agent. The AI agents operate in three basic steps: Perception, Reasoning, and Action. This paradigm is most useful for IoT. For instance, in a home security system, the perception refers to face recognition of who is latching the key to the lock, reasoning is to ensure that the face is an authorized human or otherwise and the action part is to raise notifications when the lock was opened and by whom and if required raise an alarm.

AI has an interesting role to play in project management by giving advice on basic user requirements, automating small routine tasks, and helping to predict market trends. This could enable product managers to develop better intuition, improve stakeholder engagement, and do task prioritization. An experienced product manager can sense a potential schedule slip and plan alternatives to take timely actions, such as roadmap adjustments or team composition changes. It should be mentioned here that the current AI technologies cannot yet replace an experienced project manager but can serve as an aid.

Currently, a popular AI tool is ChatGPT. It is an artificial intelligence chat bot developed by OpenAI [14] (now owned by Microsoft) and launched in November 2022. It uses a large language model (LLM) and generative AI for pattern and structure recognition to generate or compose content. OpenAI has trained it using both supervised and reinforcement learning techniques [15]. An example query and response to it have been reproduced below. Note the response to a question about the latest trends in Cloud and IoT project management.

12.8.1 What Are the Latest Trends in Cloud and IoT Project Management?

"Project management in Cloud and Internet of Things (IoT) sectors is evolving rapidly, influenced by technological advancements and emerging methodologies. Key trends include:

1. ***Integration of Artificial Intelligence (AI) and Machine Learning (ML):** AI and ML are increasingly embedded in project management tools to enhance efficiency and decision-making. These technologies enable predictive analytics, automating tasks such as resource leveling and project plan updates, thereby allowing managers to anticipate potential challenges and optimize workflows.*
2. ***Adoption of Cloud-Based Project Management Tools:** The shift towards cloud computing has transformed project management by offering flexible, scalable, and accessible tools. Cloud-based platforms facilitate real-time collaboration, centralized information access, and seamless integration across distributed teams, enhancing overall project efficiency.*
3. ***Emphasis on Cybersecurity:** With the increasing reliance on cloud-based tools and IoT devices, ensuring robust cybersecurity measures has become paramount. Project managers are prioritizing data protection by implementing multi-factor authentication, data encryption, and compliance with stringent regulations to safeguard sensitive information.*
4. ***Integration of IoT for Real-Time Data Monitoring:** IoT devices are being utilized to provide real-time data on various project parameters, such as equipment status and environmental conditions. This integration enables proactive issue resolution and informed decision-making, leading to improved project outcomes.*
5. ***Emergence of Fog Computing:** To address latency and bandwidth challenges associated with cloud computing, fog computing has emerged as a solution. By processing data closer to the source, fog computing enhances real-time analytics and responsiveness, which is particularly beneficial for IoT-intensive projects.*
6. ***Focus on Sustainability and Social Responsibility:** There is a growing trend towards incorporating sustainability and social responsibility into project*

management practices. This involves adopting eco-friendly technologies, ensuring ethical supply chains, and aligning projects with broader societal goals to meet stakeholder expectations and regulatory requirements.

These trends reflect a dynamic shift in project management approaches within the Cloud and IoT sectors, driven by technological innovation and an increased focus on security, efficiency, and sustainability."

It ought to be obvious that ChatGPT is able to respond to high-level queries and provide useful information to support decision-making. Its initial usage for project management is limited to generating ideas and creating narratives that an experienced manager may either know or can easily understand and expand upon. However, as we can see, the above response totally missed out by not mentioning Agile-based project management techniques, which have been adequately covered in the earlier chapters of this book. Thus, a good use of ChatGPT would be to use it for meaningful summarization which is a key management challenge and service.

One of the concerns using ChatGPT has been that it appears to discriminate against race and gender, which its creators are trying to mitigate. The source of this could be stereotyping and statistical bias in the input data used to train ChatGPT. Other concerns include its limited knowledge, which of course is growing with every user interaction on a daily basis. Other limitations relate to the fact that ChatGPT accepts input in the text form only and doesn't provide in-depth information, lacks expression, and cannot solve complex mathematical questions with accuracy. Despite these limitations, the tool is still useful for cross-checking and validating ideas.

Depending on how an AI model is trained in the future, it has the potential to give good initial estimates on the time and effort required to deliver a new feature, based on a team's past performance and lines of code it took to deliver similar features in the existing products. Such a model has the potential to continuously improve based on the actual result versus its predictions.

12.9 Domain-Specific SLMs

Small language models (SLMs) are AI models that can run on everyday devices like smartphones and tablets. They are designed to be more efficient and cost-effective than large language models (LLMs) for very specific and limited application context [16]. The specialized nature of SLMs can lead to improved performance and efficiency in the targeted applications. SLMs can be customized for a particular domain (e.g., legal-brief preparation) or tasks. Customization enables companies to create SLMs that are highly effective for their specific needs, such as sentiment analysis or domain-specific question answering.

Examples of SLMs include:

- Gemma: A series of SLMs from Google that are designed to be more user-friendly [17].
- Llama 3: An open-source model from Meta that is designed for AI research [18],

- Phi-3 Mini: A SLM with 3.8 billion parameters that can be deployed on smartphones [19].

Training for SLM involves three main techniques [20]:

1. **Knowledge distillation:** Here, a smaller model, or the "student," learns from a bigger, already-trained LLM, the "teacher." SLM learns smarter by focusing on refined insights from the LLM, which speeds up its learning process significantly, as compared to processing the raw data.
2. **Pruning:** This involves removing extra bits that aren't needed. By cutting out the excess parts, SLM becomes faster and leaner, which enables quicker answers from the mobile apps.
3. **Quantization:** This involves collapsing details in the data that a model learns from. This helps SLM to become lighter for running it on less powerful devices.

Project management for SLMs requires a focused approach with the following steps:

1. Defining clear goals.
2. Selecting the right data for training.
3. Fine-tuning the model for specific tasks.
4. Deploying it on appropriate hardware.
5. Continuously monitoring performance to ensure it remains accurate and efficient within the designated domain.

In addition, factors such as cost-effectiveness, data privacy, and rapid deployment capabilities unique to SLMs need to be considered for effective project execution.

12.10 Products Using SLMs

As of this writing, there are no major SLM-based products or applications in the commercial space. Perhaps one of the reasons is that even though SLMs are smaller in size compared to LLMs, the former still requires around a few billion parameters which need a lot of memory to store the model for inference in the field, which is unsuitable for many IoT products. However, we do expect a plethora of products using the aggregation of tiny SLMs in the near future. So, what then is a reasonable solution to realize this vision?

One possibility is to use Tiny Small Language Models (TSML) similar to how chip design benefited from Very Large-Scale Integration (VLSI) technology as the lithography feature size decreased over time [21]. Recently, a paper appeared explaining TinyLlama [22], which is an open-source small language model. Instead of training compute-optimal language models, this paper highlighted the importance of the inference budget, which refers to the cost of processing time and effort. This is achieved as follows. An inference-optimal language model aims for optimal performance within specific inference constraints. It demonstrated that smaller models, when trained with more data can match or even outperform their larger counterparts. This is achieved by the pre-training process of TinyLlama, using a blend of natural language and code data. This model is based on a decoder-only transformer [15]. Speed optimizations during training are obtained using Fully Sharded Data Parallel (FDSP) technique [21]. The new model is named TinyLlama v1.1, It has several key modifications as listed below:

- The team expanded beyond the singular pre-training phase of the original model and introduced a three-stage pre-training process as shown in Fig. 12.9. This includes basic pre-training, continual pre-training targeted at specific domains, and a cooldown phase.
- To reduce communication overhead, the team sharded the model parameters within a node in FSDP. It trained with a cluster consisting of 16 nodes, each equipped with four A100 Nvidia Ampere 40GB GPUs, and set the batch size to approximately 1.8 million tokens.
- The team reduced the total number of pre-training tokens from 3 to 2 trillion. Despite the reduction in training data volume, only a marginal improvement in performance on downstream tasks was observed.

The resulting model was evaluated on a wide range of commonsense reasoning and problem-solving tasks showing that TinyLlama can enable end-user applications on mobile devices and other lightweight edge computing platforms [22].

The top use cases [23] of SLMs include the following domains:

1. Mobile applications: SLMs are used for on-device language processing tasks to execute on mobile devices for applications such as investment banking.

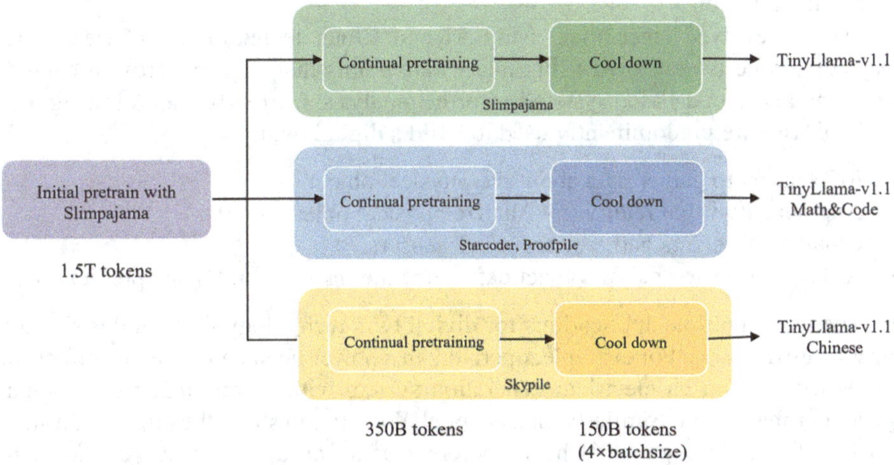

Fig. 12.9 Pre-training stages and specialization pipeline of TinyLlama [22]

2. IoT and Edge Devices: SLMs empower IoT devices with natural language interfaces and intelligent data processing.
3. Healthcare: SLMs assist with tasks like medical transcription and initial patient assessment.
4. Education: SLMs power intelligent tutoring systems and automated grading tools. They provide personalized learning experiences and instance feedback to students.
5. Customer service: SLMs are used as chatbots and sentiment analysis. This allows for quick, automated responses and better understanding of customer needs.
6. Finance: SLMs assist with fraud detection and automated report generation. They help process large volumes of text data quickly and securely.
7. Content creation and curation: SLMs aid in content summarization, SEO optimization, and automated content generated.
8. Embedded systems: SLMs enable natural language interfaces in various devices. This allows for more intuitive human–machine interaction in products like smart appliances and vehicles.

12.11 Role of Digital Twins

Digital Twin (DT) is a digital replica of a physical entity. This entity may be used to simulate or emulate the object's state transitions [24]. DT concept was first used by the NASA space program to simulate a spacecraft in the 1960s and debug flight issues in real-time as they manifested. This concept was used during Apollo 13 when NASA simulated conditions to safely return the spacecraft and astronauts after a malfunction.

DT has evolved since then using AI and Cloud technologies. Currently, an object's profile is generated using many data points that sensors across it capture and send to a Cloud-based system for further analysis. Currently, the following four technologies are predominantly used to build a digital twin:

- IoT sensors to gather data about the physical object.
- Augmented/virtual reality to visualize physical objects in 3D.
- Cloud to store data gathered from IoT sensors.
- AI based tools to analyze object data, generate insights, and make predictions.

There are multiple applications for digital twin technology in the industrial and healthcare domains. For example, operators in a power plant can study the effects of increased load on an electrical generation system without distributing the actual generator machinery. Similarly, in a hospital, doctors can study the effects of a stent on blood flow in a patient's heart, before placing an actual stent. As shown in Fig. 12.10, this market is slated to grow at an astounding 57% Cumulative Annual Growth Rate (CAGR) [24].

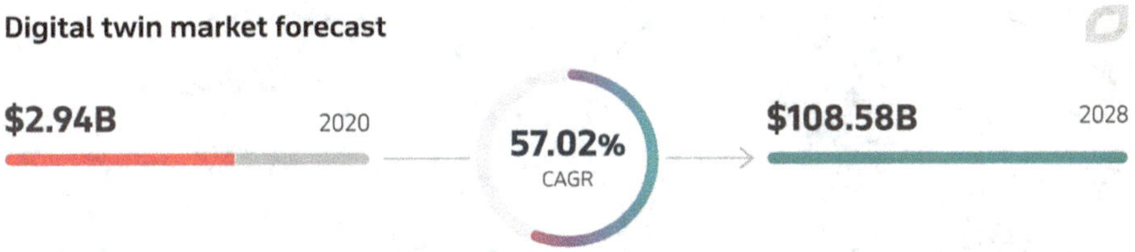

Fig. 12.10 Digital twin market size and forecast [24]

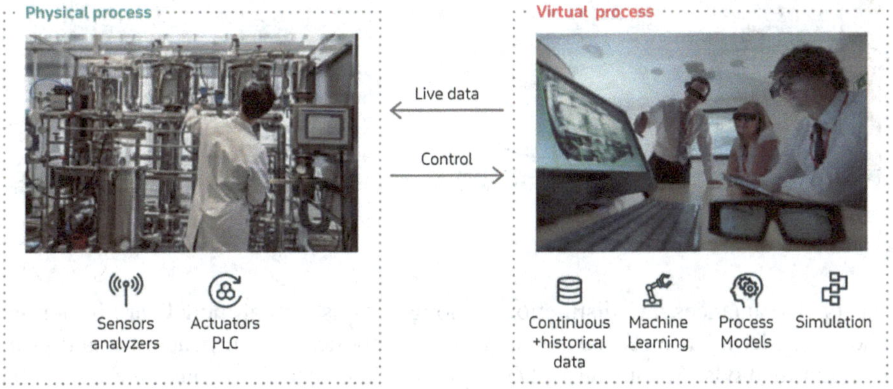

Fig. 12.11 Digital twin for pharmaceutical manufacturing [24]

One example of digital twin usage is in pharmaceutical companies [24]. Traditionally, these companies can assess a product's quality only after it has been made, implying that if it's low, the whole batch has to be discarded. This results in a waste of time and materials. Digital twins enable the assessment of product quality during production, assisting manufacturers in adjusting operations to meet quality standards, as shown in Fig. 12.11.

Other use cases for digital twins include the following [25]:

- Using predictive maintenance to maintain equipment and production lines.
- Monitoring products in real time as they are used by end users.
- Manufacturing process optimization.
- Enhancing product traceability processes.
- Testing, validating, and refining assumptions.
- Increasing the level of integration between unconnected systems.
- Remote troubleshooting of equipment, regardless of geographical location.

In the field of healthcare, as shown in Fig. 12.12, interactions between a physical entity (i.e., a patient), and its corresponding digital twin (virtual patient) could traverse from microscopic to macroscopic scale and span the entire lifespan of individuals [23]. In the context of a patient-specific DT, prediction and analysis from DT can be used to help the patient with better treatment outcomes and less adverse

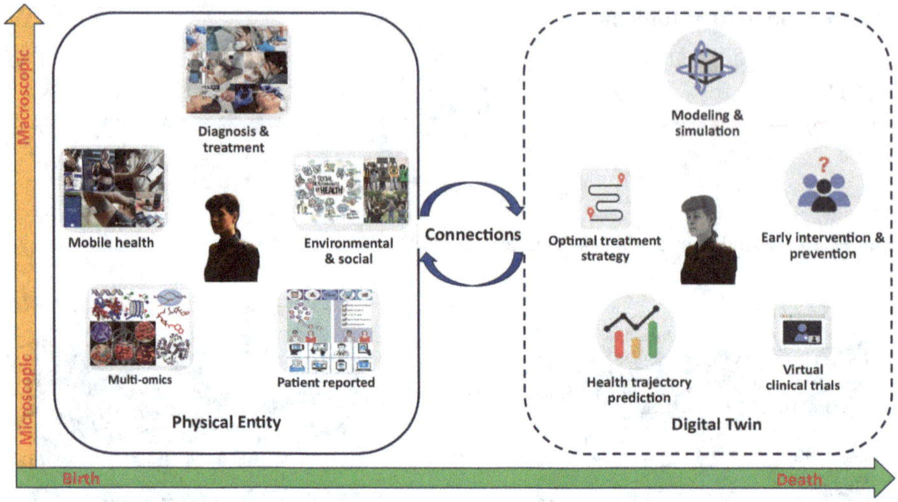

Fig. 12.12 Digital twin for healthcare [26]

effects. The nuances and distinctions among various virtual models are important when considering aspects such as data sources, applications, interactions, and visualization methods. For instance, DTs that simulate specific organs, like the living heart model, are mainly developed using detailed imaging data. This is in contrast with disease models aimed at supporting precision medicine, which includes molecular profiling data and clinical information.

The applications of these models are as diverse as their data sources. Organ-specific digital twins, such as those replicating heart function, help in predicting the performance of mechanical medical devices such as pacemakers. On the other hand, disease models provide invaluable insights into the effectiveness of pharmaceutical interventions, especially in how drugs interact with complex biological processes [26].

12.12 An AI-Powered "GPS for Learning"

Complex human experiences—whether in learning, healthcare, or financial decisions—are difficult to fit into a standard model. This is because their outcomes emerge through countless non-linear interactions between interconnected dimensions. Traditional approaches attempt to segment individuals into groups for fairly standardized patterns of experiences, but this fundamentally fails to address unique individual needs, particularly in education. There are two primary reasons for it:

1. Learning pathways depend on each person's specific context, such as prior knowledge.
2. Each person has a different rate of learning.

MyGooru [27] has implemented AI-powered "GPS for Learning" that guides each learner through personalized pathways to their learning goals. Just as Google Maps transforms vast amounts of information about roads, public transportation, traffic conditions, businesses, and accidents into a personalized journey all about you—where you are, where you want to go, and your preferences—MyGooru does the same for learning. As shown in Fig. 12.13, Gooru's GPS for learning factors in each student's current knowledge, interests, abilities, and preferred learning approaches to plan a path to reach the chosen goals.

MyGooru's App empowers learners in schools, colleges, corporations, and lifelong learners with AI-driven personalized, competency-based learning. MyGooru identifies knowledge gaps and tracks progress in real time, recommending personalized learning pathways and activities. It provides a GPS-like experience that guides learners to their learning destination with activities appropriate to their unique knowledge, skills, and mindsets. It starts by providing a personalized pathway of learning activities and subsequently nudges the learner based on their performance. MyGooru App includes curated courses across K12, higher education, skills training, and professional learning.

Whereas GenAI platforms treat learners as collections of independent attributes, MyGooru recognizes learners as complex systems with deeply interconnected dimensions. Standard LLMs and agents can generate personalized content based on pattern matching from the training data, but they fundamentally lack the mathematical framework to understand how progress cascades across different aspects of learning.

Fig. 12.13 MyGooru's AI power "GPS for Learning" system [27]

Addressing complex human experiences requires sophisticated mathematical frameworks beyond traditional AI models. Gooru has developed the polyline model [28], a mathematical foundation specifically designed to represent and operate complex systems. The polyline model formally defines a polyline as a finite sequence of connected line segments in a multi-dimensional space, where each segment represents elements within domain understanding. This enables precise mathematical operations through polyline algebra, including fundamental operations like union (∪), intersection (∩), and complement (¬) that can be composed into sophisticated derivations. As shown in Fig. 12.14, these operations enable the system to compute valid state transitions while preserving relationship networks, calculate how changes in one dimension affect related attributes, and maintain system coherence as new patterns emerge.

Unlike traditional approaches that rely on fixed attributes with independent values, the polyline model represents each person's state as probability distributions across interconnected dimensions that evolve through their unique interactions. This system works through an agentic workflow called ALCMI (Aggregate, Locate, Curate, Mediate, Integrate) that orchestrates personalization:

1. **Aggregate:** Processes all available learning data to build a comprehensive understanding.
2. **Locate:** Identifies the learner's current state and goals across multiple dimensions.
3. **Curate**: Assembles personalized resources and recommendations.
4. **Mediate:** Generates optimal pathways towards learning destinations.
5. **Integrate:** Incorporates responses to continuously refine guidance.

MyGooru's workflow, as shown in Fig. 12.15, is powered by the personalization transformer, which enhances large language models with mathematical constraints

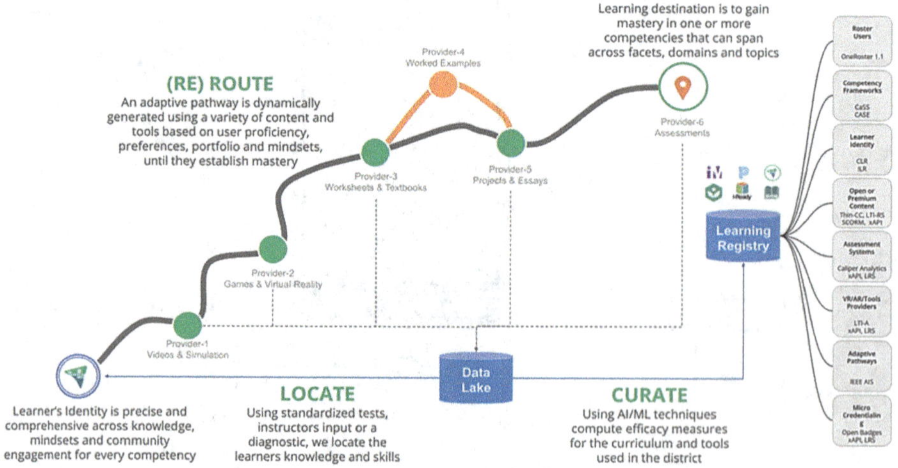

Fig. 12.14 Polyline model for navigator AI technology [28]

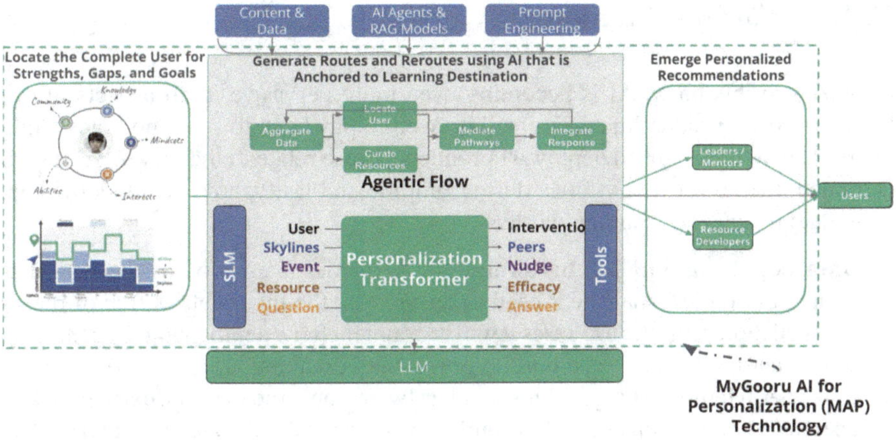

Fig. 12.15 MyGooru Workflow for learning

to ensure each recommendation contributes meaningfully to the learner's journey while adapting to their unique context and needs.

MyGooru's App development required progress across three parallel tracks:

1. Mathematical Research: Advancing theoretical foundations.
2. Technology Development: Implementing theoretical concepts in scalable systems.
3. Field Implementation: Validating approaches in authentic contexts.

MyGooru's project managers had to balance progress across these interconnected but distinct workstreams, ensuring that advances in one area informed and supported the others.

MyGooru's approach has been validated through 14 years of research and implementations with 5.3 million users globally. Results demonstrate unprecedented improvements in both learning outcomes and business metrics [29]:

- 2.8× acceleration in learning growth in K-12 implementations
- 30% increase in engagement for commercial platforms
- 25% improvement in trial-to-paid conversion rates
- 15% enhancement in retention metrics

These results have been observed [30] across diverse implementations, from K-12 school districts to professional training programs, demonstrating the universal applicability of the approach across learning contexts.

We would like to point out that an Education Technology (EdTech) product such as MyGooru affords two different forms of licensing, for value creation:

1. **Product licensing:** Direct implementation by educational institutions (100+ implementations).
2. **Technology licensing:** Strategic partnerships with major platforms (Pearson, Barnes & Noble, PowerSchool, Stemuli).

12.13 Words of Caution

As we close this book, AI is becoming even more pervasive in all aspects of our lives. However, product managers will do well to pay heed to some potholes along the journey to develop and deploy AI products. These relate to the way an AI model is trained, the power it consumes during runtime and lastly, how the model is kept up to date in an ever-changing world.

1. **Data poisoning and jail breaking:** An AI model is as good as the data it is trained on. Project managers need to ensure that the input data is free of biases and well represents the use cases it will encounter after a deployment. Specifically, some guard rails need to be built to prevent jail breaking [15].
2. **Power efficiency:** AI algorithms and hardware consume extraordinary power to operate. Cloud companies often build power plants or locate their datacenters near existing power plants to meet the AI demands of scalability [1]. Project managers need to be cognizant of low power constraints for any Edge-based devices [15] and ensure that their products can operate within the given power specifications.
3. **Continuous model updates:** Since the real world around us, and its resulting data in Internet servers, is changing continuously, any training done for an AI model gets out of date the moment it is created. Tesla has a method to update software in cars using Over the Air (OTA) methods. Similarly, project managers need to plan for keeping their AI model updated in the field.

12.14 Summary

Cloud is a double-edged sword and can be disruptive if not handled properly. It has the power to upend existing businesses, products, and use cases. Recall when a fax machine posed challenges to FedEx's overnight shipping business for urgent document sharing the latter adapted by expanding its infrastructure to send large packages and survived this challenge. Similarly, Cloud challenges how data and information are exchanged between the stakeholders. However, Cloud itself needs to evolve by adding security features, such as encryption so that mission critical data can be sent over public networks reliably and securely. Developing new applications or migrating existing solutions to the Cloud presents opportunities for those who adapt. We studied Large Language Models (LLMs) and potential usage of domain-specific Small Language Models (SLMs). We examine the role and market forecasts for digital twins, and their relations to IoT-based products. We conclude this chapter with an AI-based Education Technology (EdTech) application, MyGooru.

Points to Ponder

1. Think of Cloud-based applications installed on your smart phone. How many of these are supported by a common service provider, versus hosted in different datacenters?
2. Even if applications belong to different companies, their services may be hosted by a common Cloud service provider. Can you think of an example?
3. Why is data quality very important for training SLM?
4. How can digital twins assist with personalized medicine?
5. What can go wrong with digital twins and how can this be prevented?

Answers

1. Think of Cloud-based applications installed on your smart phone. How many of these are supported by a common service provider, versus hosted in different datacenters?

   (a) A typical phone user may have dozens of app, including emails, games, financial, and social networking. Chances are that all of these are being supported by different service providers (e.g., Google, Apple, or Microsoft) and being hosted in different datacenters. Sometimes, even the same service provider uses different datacenters, e.g., Google has multiple locations to serve its worldwide customers to minimize latency.

2. Even if applications belong to different companies, their services may be hosted by a common Cloud Service Provider. Can you think of an example?

   (a) Sometimes two different application providers may share a common Cloud service provider, such as Netflix and Salesforce, both of which are hosted by Amazon's AWS.

3. Why is data quality very important for training SLM?

   (a) An AI model performs as well as the data it was trained on, so quality of training data is important for both LLMs and SLMs.
   (b) However, SLMs are domain specific and not as big or complex (in terms of parameters) as LLMs, the former relies on the quality of data they were trained on.
   (c) Imagine application of SLM in a specialized area, such as preparing legal briefs. A SLM trained on legal documents is expected to do a much better than one learning from random Internet pages. The same goes for healthcare, SLMs trained on accurate medical information can really help doctors make better decisions because they're getting suggestions that are informed by reliable data.

4. How can digital twins assist with personalized medicine?

   (a) Medicine is practiced by adhering to established protocols developed over time by medical professionals based on field data.

(b) It often disregards individual genetic makeup or personal situations; for example, no two cancers are alike, yet chemo or radiation follows a standardized plan.
(c) A properly constructed Digital Twin (DT) model can forecast a patient's response to treatment, thereby assisting physicians in personalizing medical care.

5. What can go wrong with digital twins and how to prevent it?

(a) A Digital Twin (DT) is an AI/ML-based model that is as good as the training data used to build it.
(b) Modeling complex systems, such as the human body, weather, or stock markets, proves to be challenging even with the use of billions of parameters.
(c) The quality of the DT will depend on the quality of IoT sensors and products used.
(d) Differing results between reality and the model's prediction can be used to improve the DT.

References

1. Sehgal, N. K., Bhatt, P. C. P., & Acken, J. M. (2020). *Cloud computing with security and scalability*. Springer. https://link.springer.com/book/10.1007/978-3-031-07242-0
2. https://www.j2interactive.com/blog/brief-history-salesforce
3. https://en.wikipedia.org/wiki/Microsoft_365
4. https://bstrategyhub.com/netflix-business-model-how-does-netflix-make-money/
5. https://en.wikipedia.org/wiki/List_of_largest_Internet_companies
6. https://en.wikipedia.org/wiki/Salesforce
7. https://architect.salesforce.com/
8. https://www.a10networks.com/blog/microsoft-office-365-enterprise-network-architecture/
9. https://www.geekwire.com/2022/microsoft-revenue-growth-slows-as-cloud-business-helps-offset-windows-decline/
10. https://support.microsoft.com/en-us/office/collaborate-from-anywhere-using-microsoft-365-ac05a41e-0b49-4420-9ebc-190ee4e744f4
11. https://www.codingninjas.com/codestudio/library/system-design-netflix-a-complete-architecture
12. https://microservices.io/
13. Bahssas, D. M., et al. (2015). Enterprise resource planning (ERP) systems: Design, trends and deployment. *The International Technology Management Review, 5*(2), 72–81. https://www.atlantis-press.com/journals/itmr/24089
14. https://openai.com/
15. Sehgal, N., Saxena, M., & Shah, D. (2024). *AI on the edge with security*. https://link.springer.com/book/10.1007/978-3-031-78272-5
16. https://venturebeat.com/ai/why-small-language-models-are-the-next-big-thing-in-ai/
17. https://ai.google.dev/gemma
18. https://ai.meta.com/blog/meta-llama-3/
19. https://huggingface.co/microsoft/Phi-3-mini-128k-instruct
20. https://www.superannotate.com/blog/small-language-models
21. https://en.wikipedia.org/wiki/Very-large-scale_integration

22. https://arxiv.org/pdf/2401.02385
23. https://kanerika.com/blogs/small-language-models/
24. https://atos.net/wp-content/uploads/2022/07/Atos_Manufacturing_Computer_vision.pdf
25. https://www.itransition.com/blog/digital-twin-manufacturing
26. Katsoulakis, E., Wang, Q., Wu, H., et al. (2024). Digital twins for health: A scoping review. *Npj Digital Medicine, 7*, 77. https://doi.org/10.1038/s41746-024-01073-0
27. https://mygooru.ai/
28. https://mygooru.ai/technology
29. https://onlinelibrary.wiley.com/doi/pdf/10.1002/hbe2.169
30. https://www.christenseninstitute.org/publications/connecting-ed-and-tech/

Index

P. C. P. Bhatt et al., *Project Management in Cloud and IoT Applications*,
https://doi.org/10.1007/978-3-031-99427-2